# Fiber-Reinforced Composites - Recent Advances, New Perspectives and Applications

*Edited by Longbiao Li*

Published in London, United Kingdom

Fiber-Reinforced Composites - Recent Advances, New Perspectives and Applications
http://dx.doi.org/10.5772/intechopen.1001621
Edited by Longbiao Li

Contributors
Ahmet Hayrullah Sevinç, Baosheng Xu, Chunmei Chen, Dmytro Tiniakov, Dorys C. González, Hayriye Hale Aygün, Huanbo Wang, Hui Wang, Jesús Mínguez, Kaimeng Xu, Longbiao Li, Miguel A. Vicente, Muhammed Yasin Durgun, Ni Liu, Runze Jin, Ryan Randall, Sattar Mohammadi Esfarjani, Tian Liu, Álvaro Mena-Alonso



First published in London, United Kingdom, 2024 by IntechOpen
IntechOpen is the global imprint of INTECHOPEN LIMITED, registered in England and Wales, registration number: 11086078, 167-169 Great Portland Street, London, W1W 5PF, United Kingdom

British Library Cataloguing-in-Publication Data
A catalogue record for this book is available from the British Library

Additional hard and PDF copies can be obtained from orders@intechopen.com

Fiber-Reinforced Composites - Recent Advances, New Perspectives and Applications
Edited by Longbiao Li
p. cm.
Print ISBN 978-0-85466-758-1
Online ISBN 978-0-85466-757-4
eBook (PDF) ISBN 978-0-85466-759-8

For EU product safety concerns:
IN TECH d.o.o., Prolaz Marije Krucifikse Kozulić 3, 51000 Rijeka, Croatia,
info@intechopen.com or visit our website at intechopen.com.

# Meet the editor

Dr. Longbiao Li is a lecturer in the College of Civil Aviation at the Nanjing University of Aeronautics and Astronautics. Dr. Li's research focuses on the vibration, fatigue, damage, fracture, reliability, safety, and durability of aircraft and aero engines. In this research area, he has published more than 200 SCI journal publications, 12 monographs, 4 edited books, 4 textbooks, 3 book chapters, 34 Chinese patents, 2 U.S. patents, 6 Chinese software copyrights, and more than 30 refereed conference proceedings. He has been involved in different projects related to structural damage, reliability, and airworthiness design for aircraft and aero engines, supported by the Natural Science Foundation of China, COMAC Company, and AECC Commercial Aircraft Engine Company.

# Contents

# Preface

Fiber-reinforced composites possess high specific strength and modulus and have been widely applied in aerospace, automotive, and nuclear industries. To ensure operation reliability and safety, it is necessary to conduct damage monitoring methods and develop models, approaches, and tools to predict the mechanical behavior of the fiber-reinforced composites in different operation conditions. The book focuses on recent advances, new perspectives, and applications of different fiber-reinforced composites, and examples of engineering applications have been illustrated, which can help the readers better understand the design, manufacturing, and operation of fiber-reinforced composites.

**Longbiao Li**
College of Civil Aviation,
Nanjing University of Aeronautics and Astronautics,
Nanjing, P.R. China

Chapter 1

# Progress on Application of Ceramic-Matrix Composites in Hot-Section Components in Aero Engines

*Longbiao Li*

## Abstract

Ceramic-matrix composites (CMCs) are a class of materials that combine the high-temperature stability and strength of ceramics with the toughness and damage tolerance of fibers. One of the key advantages of CMCs is their ability to withstand high temperatures, making them ideal for applications in gas turbines, rocket nozzles, and heat exchangers. The high-temperature stability of CMCs is due to the ceramic-matrix material, which has a high melting point and excellent thermal conductivity. This allows CMCs to operate at temperatures above 1000°C, where traditional metal alloys would fail. These composites have gained significant attention in recent years due to their potential for use in a wide range of applications, including aerospace, automotive, energy, and defense. This chapter reviews the recent progress in the design, application, and challenges in CMCs, especially the hot-section components design and testing at elevated temperatures.

**Keywords:** ceramic-matrix composites, hot-section components, aero engine, combustor liner, turbine, exhaust components

## 1. Introduction

Ceramic-matrix composites (CMCs) with continuous fiber reinforcement possess high specific strength and modulus, and have already been recognized as attractive structural ceramic materials for gas turbine hot-section components in civil and military aero engines [1, 2]. The tensile curves of CMCs show obvious nonlinear behavior due to multiple damage mechanisms, e.g., matrix cracking, interface debonding, and fiber failure and pullout [3]. Compared to monolithic ceramics, CMCs can be used under high thermos-mechanical loads [4–6].

This chapter presented a review on the recent design, application, and challenge of CMCs in hot-section components in aero engines, e.g., combustor liners, turbine components, and exhaust components. These hot-section components were developed by France, United States, China, Japan, etc., and have already been applied in military or commercial aero engines.

## 2. Application of ceramic-matrix composites in aero engines

### 2.1 Application on combustor liner

In the early 1980s, SNECMA company started research on the application of CMCs in hot-section components of aircraft engines. The CERASEPR series CMC materials were developed using chemical vapor infiltration (CVI) technology and tested on M88 engines. On this basis, SNECMA has upgraded and improved the CERASEPR series materials, as well as the high-temperature mechanical properties and temperature resistance of the materials. SNECMA company used improved materials to produce a full-size combustion chamber component [7], as shown in **Figure 1**. The CERASEP®-A410 afterburner flame holders were tested for 143 hours with a maximum temperature of 1180°C without any defect detected by nondestructive evaluation (NDE), and the CERASEP®-A415 combustor liner was tested for 180 hours (600 representative cycles) including 100 hours of maximum working conditions. The combustor linear with the CMC mixer was demonstrated on CFM56 engine. The weight gain of the CMC mixer was about 35%.

GE and Allison collaborated to develop and verify the CMC combustor liners on XTC76/3 and passed a 5000 hrs full life-cycle operation test and 500 h high-temperature operation test at 1200°C [8], as shown in **Figure 2**.

GE has manufactured and validated a SiC/SiC low-emission combustor liner with multiple inclined cooling hole [9], as shown in **Figure 3**. Experiments have shown that using this SiC/SiC combustor liner can reduce the cooling air of the combustor by 50%, reduce the mass by 50%, and reduce the $NO_x$ emissions by about 20%. The CMC inner and outer combustor liners, high-pressure turbine (HPT) stage 1 and stage 2 nozzles, and stage 1 shrouds were expanded to the GE9X aero engine.

UTRC (United Technologies Research Center) and P&W Canada validated the SiC/SiC combustor in a PW 200 series combustor [10], as shown in **Figure 4**. The fill annular CMC combustor rig was engine tested for 250 cycles between idle (40,000 rpm) and full power (57,000 rpm) and severely tested the response of the CMC/metal interfaces to accelerated thermal cycling. The test was stopped after 250 cycles, with no damage observed.

**Figure 1.**
*CERASEP®A415 rig and inner & outer liners of an engine combustor [7].*

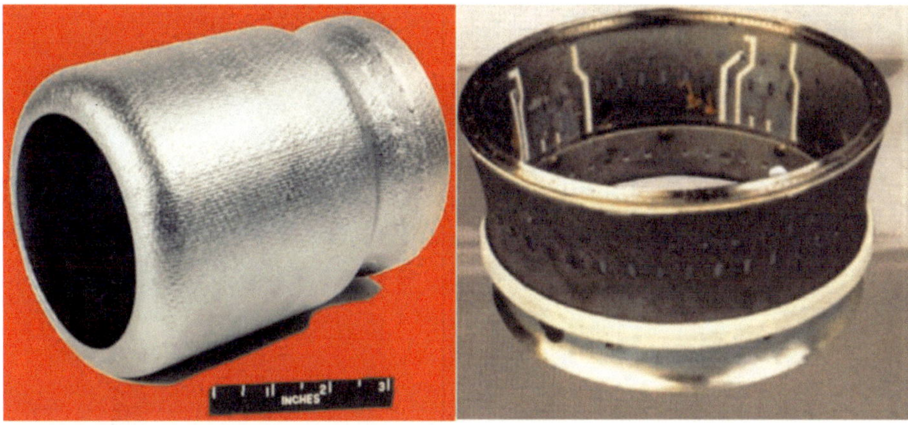


**Figure 2.**
*CMC combustor liner by GE and Allison [8].*

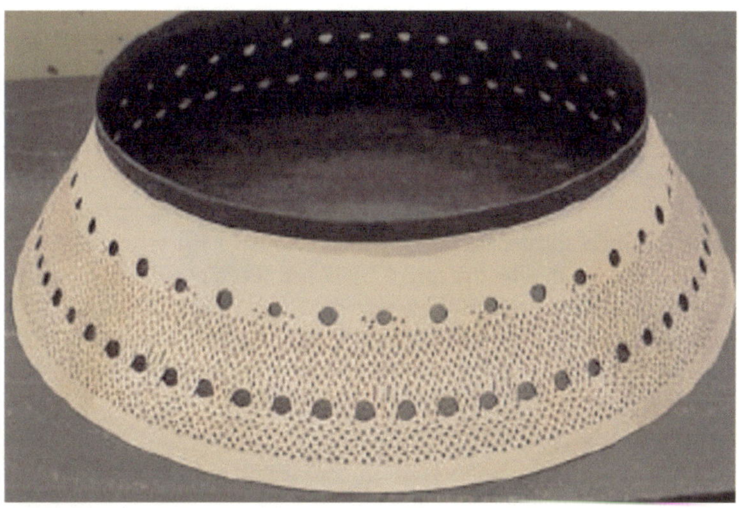

**Figure 3.**
*Low emission SiC/SiC combustor liner developed by GE company.*

**Figure 4.**
*MI SiC/SiC combustor before and after rig test [10].*

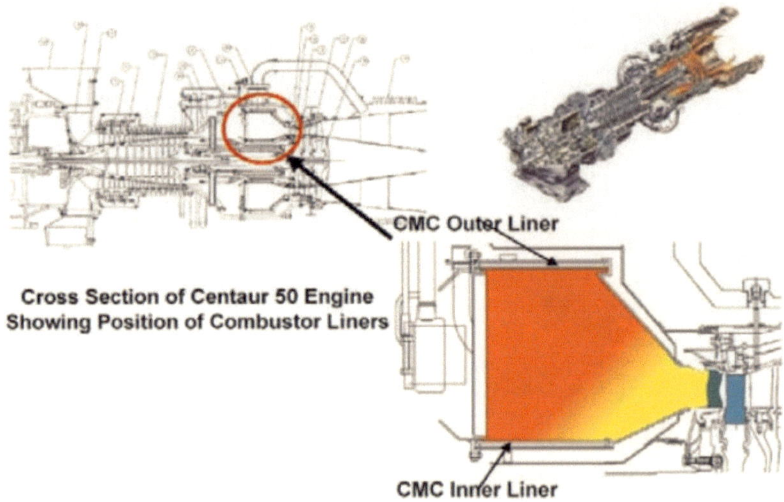


**Figure 5.**
*The SiC/SiC inner linear and oxide/oxide outer liner tested on Solar Centaur 50S combustor [11].*

Solar Turbines Incorporated developed and evaluated both SiC/SiC and oxide/oxide combustor liners in test rigs and Solar Centaur® 50S engines since 1992 [11], as shown in **Figure 5**. The development roadmap includes the rig testing of subscale combustors, full-scale liner tests in atmospheric and high-pressure combustor rigs, and in-house and field testing in actual production engines, as shown in **Figure 6**. **Table 1** lists the engine field test of the CMC combustor liner on the Centaur® 50S engine. Test durations of 15,144 hrs and 13,937 hrs were conducted on the SiC/SiC liners with protective environmental barrier coatings (EBCs), and 12,582 hrs were conducted for oxide/oxide liner with a Friable Graded Insulation (FGI) coating. $NO_x$ and CO emissions were < 15 ppmv and < 10 ppmv, respectively.

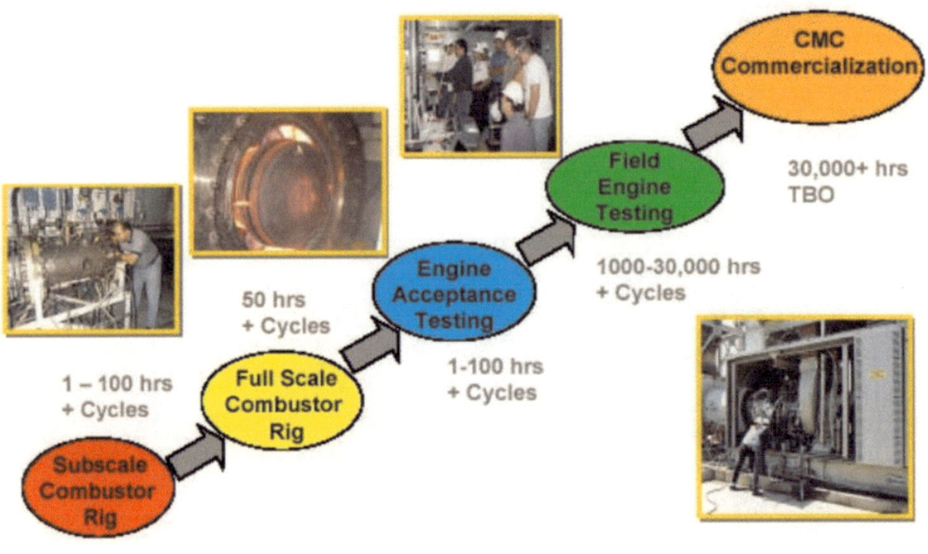


**Figure 6.**
*Solar's CMC combustor liner development roadmap [11].*

| Engine field test | Materials | Interphase | Fabrication method | EBCs | Test duration (h)/ starts | Maximum temperature/ (°C) | Configuration of the liner | Failure mechanisms |
|---|---|---|---|---|---|---|---|---|
| CT-1 | Ceramic Grade Nicalon SiC fiber reinforced enhanced-SiC | PyC | Chemical Vapor Infiltrated (CVI) | NONE | 948 h (15 starts) | 1260 |  Inner and outer liner | Environmental degradation attributed to accelerated oxidation of SiC due to the presence of $H_2O$ in the combustion environments. 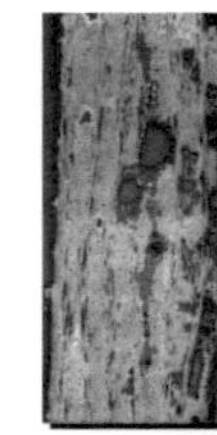 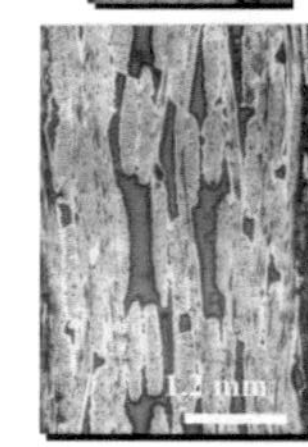 |
| CT-2 | Hi-Nicalon SiC fiber reinforced enhanced-SiC | BN | Chemical Vapor Infiltrated (CVI) | NONE | 352 h | 1175 | Inner liner | Due to the failure of the AS800 turbine blades, the test was stopped. |
| | Hi-Nicalon SiC fiber reinforced enhanced-SiC | PyC | | | | | Outer liner | |

| Engine field test | Materials | Interphase | Fabrication method | EBCs | Test duration (h)/ starts | Maximum temperature/ (°C) | Configuration of the liner | Failure mechanisms |
|---|---|---|---|---|---|---|---|---|
| CT-3 | Hi-Nicalon SiC fiber reinforced enhanced-SiC | BN | Chemical Vapor Infiltrated (CVI) | NONE | 1906 h | 1175 | Inner liner | Surface oxidation at the fuel injector impingement locations, the maximum recession approached approximately 0.2 mm. |
| | Hi-Nicalon SiC fiber reinforced enhanced-SiC | PyC | | | | | Outer liner | Surface oxidation at the fuel injector impingement locations |
| CT-4 | Hi-Nicalon SiC fiber reinforced SiC+Si | BN | Melt-infiltrated (MI) | NONE | 2758 h (26 starts) | 1260 | Inner liner | Better condition than both the CVI inner and outer liners. |
| | Hi-Nicalon fiber reinforced enhanced-SiC | PyC | Chemical Vapor Infiltrated (CVI) | | | | Outer liner | Heavily degraded, having lost about 80% of the wall thickness, and the residual strength of the liner was about 50% of the original strength at room temperature and 1200°C. |
| CT-5 | Hi-Nicalon SiC fiber reinforced SiC+Si | BN | Melt-infiltrated (MI) | Si/mullite/BSAS 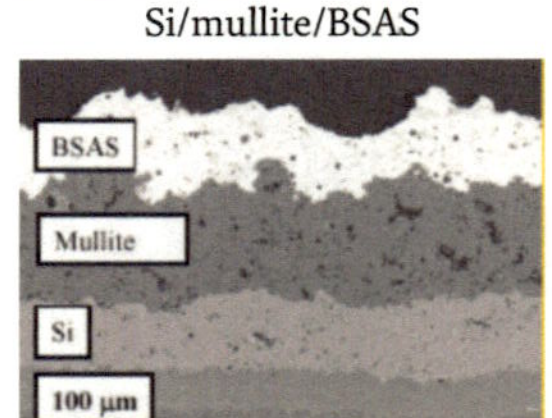  | 13,937 h (61 starts) | | Inner 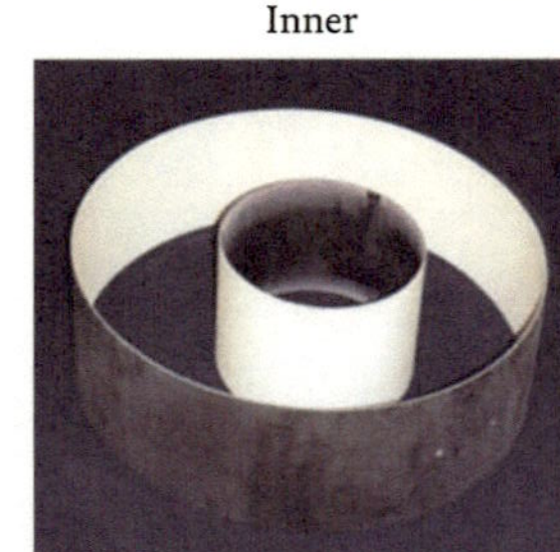 | Heavy degradation of the EBC and of the underlying CMCs in areas where the EBC had debonded |

| Engine field test | Materials | Interphase | Fabrication method | EBCs | Test duration (h)/ starts | Maximum temperature/ (°C) | Configuration of the liner | Failure mechanisms |
|---|---|---|---|---|---|---|---|---|
| | Hi-Nicalon fiber reinforced enhanced-SiC | PyC | Chemical Vapor Infiltrated (CVI) | Si/mullite + BSAS/BSAS | | | Outer 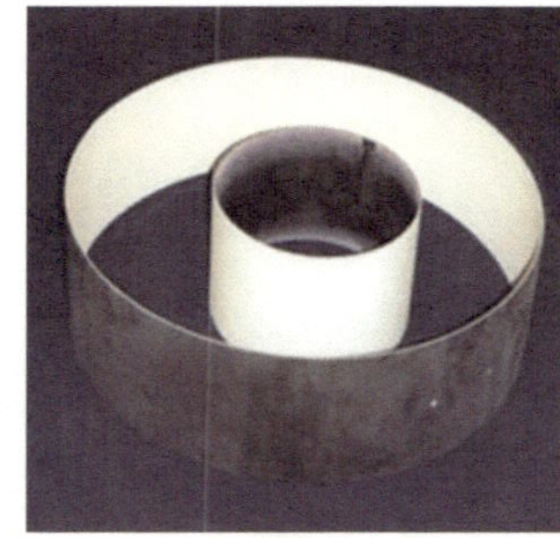 | |
| MM-1 | Hi-Nicalon SiC fiber reinforced SiC+Si | BN | Melt-infiltrated (MI) | Si/mullite + BSAS/BSAS | 7238 h (159 starts) | | Inner  | The outer liner had significantly more EBC spallation than the inner liner. |
| | Hi-Nicalon fiber reinforced enhanced-SiC | PyC | Chemical Vapor Infiltrated (CVI) | | | | Outer  | |

| Engine field test | Materials | Interphase | Fabrication method | EBCs | Test duration (h)/ starts | Maximum temperature/ (°C) | Configuration of the liner | Failure mechanisms |
|---|---|---|---|---|---|---|---|---|
| CT-6 | Hi-Nicalon SiC fiber reinforced SiC+Si | BN | Melt-infiltrated (MI) | Si/BSAS | 5135 h (43 starts) | | Inner  | There was EBC degradation for both liners. |
| | Hi-Nicalon fiber reinforced enhanced-SiC | PyC | Chemical Vapor Infiltrated (CVI) | Si/mullite + BSAS/BSAS | | | Outer  | |
| MM-2 | Tyranno ZM SiC fiber reinforced SiC+Si with SiC seal coat | BN | Melt-infiltrated (MI) | Si/mullite + BSAS/BSAS | 15,144 h (92 starts) | | Inner linear | Circumferential and lateral cracks propagated through the thickness. Extensive oxidation and surface recession on the cold-side surface of the inner liner. The presence of $SiO_2$ oxidation layer at the Si/mullite+BSAS interface. |

| Engine field test | Materials | Interphase | Fabrication method | EBCs | Test duration (h)/ starts | Maximum temperature/ (°C) | Configuration of the liner | Failure mechanisms |
|---|---|---|---|---|---|---|---|---|
| | Hi-Nicalon SiC fiber reinforced enhanced-SiC with SiC seal coat | PyC | Chemical Vapor Infiltrated (CVI) | | | Outer liner | | |
| MM-3 | Tyranno ZM SiC fiber reinforced SiC+Si with SiC seal coat | BN | Melt-infiltrated (MI) | Si/SAS | 8368 h (32 starts) | | Inner | Significant spallation of the SAS topcoat on the inner liner. The dual layer Si/SAS EBC was unstable. |
| | Tyranno ZM SiC fiber reinforced SiC+Si with SiC seal coat | BN | Melt-infiltrated (MI) | Si/mullite+SAS/SAS | | | Outer | EBC on the outer liner was in good condition. |
| CT-7 | Hi-Nicalon SiC fiber reinforced enhanced-SiC | BN | Chemical Vapor Infiltrated (CVI) | | 12,582 h (63 starts) | | Inner liner | |
| | $Al_2O_3/Al_2O_3$ | | | Friable Graded Insulation (FGI) surface coating | | 1149 | Outer liner 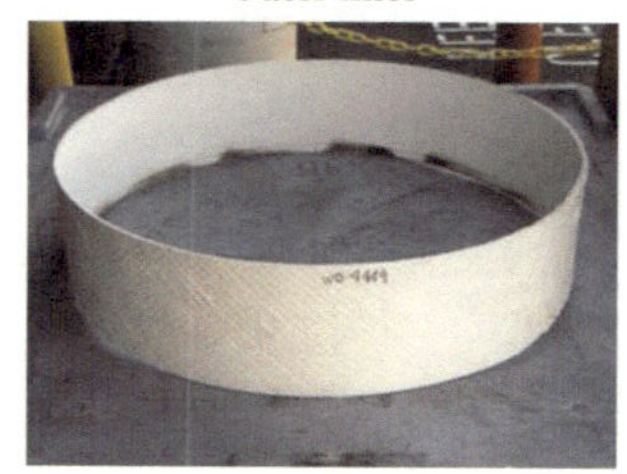 | A few loose tows at the cold (OD) side. |

**Table 1.**
*Engine field test of CMC combustor liner in centaur 50S engine.*

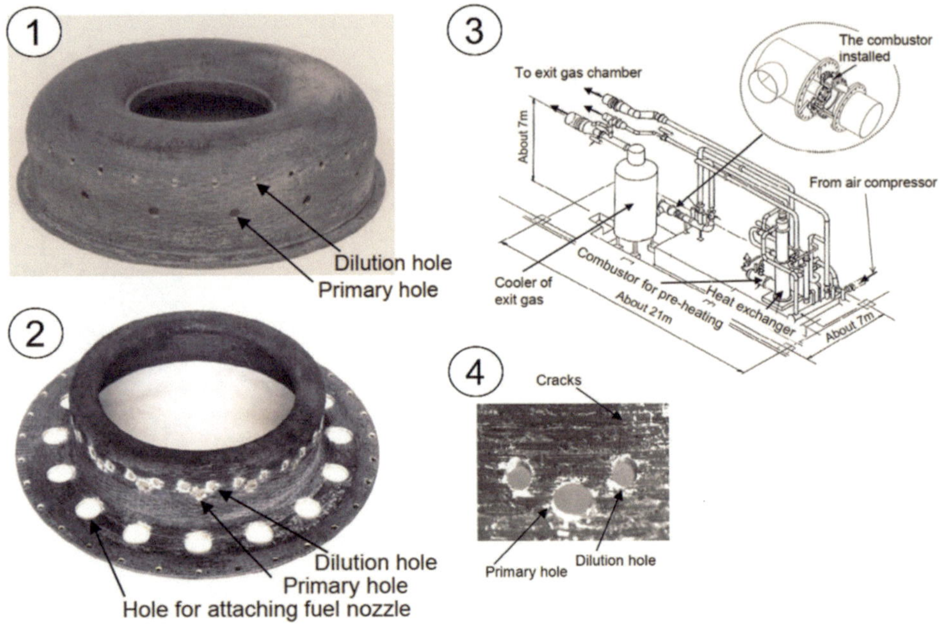


**Figure 7.**
*SiC/SiC combustor liner developed by Kawasaki heavy industries. ① SiC/SiC outer liner. ② SiC/SiC inner liner. ③ Test facility. ④ Cracks near dilution holes on the inner liner.*

The Kawasaki Heavy Industries, Ltd. developed the uncooled three-dimensional Tyranno ZMI™ SiC fiber reinforced SiC matrix composite liners using the polymer impregnation and pyrolysis (PIP) process [12, 13], as shown in **Figure 7**. The low cycle fatigue (LCF) test was carried out, and the test condition was varied periodically from the idle to the design point. 65 cycle test was finally carried out until the first detection of cracks by bore-scope inspection.

German aerospace center developed the oxide/oxide tubular combustor liner for a lean combustor in a future aero engine in the medium thrust range and tested at engine conditions [14–16], as shown in **Figure 8**. The tubular CMC combustion chamber and the EBC were in good condition.

**Figure 8.**
*Oxide/oxide combustor liner developed by German Aerospace Center [15]. ① Oxide/oxide combustor liner for component test. ② Fit check with the oxide/oxide combustor liner. ③ Before the test. ④ After the test.*

## 2.2 Application on turbine components

### *2.2.1 Turbine guide vanes*

In the United States, under NASA's Enabling Propulsion Materials and Ultra Efficient Engine Technologies programs, the SiC/SiC turbine guide vane was developed [17], as shown in **Figure 9**. The vane consisted of six plies of 2D Sylramic™ SiC fiber cloth (Y-cloth) reinforced with a CVI/slurry-cast/melt-infiltrated SiC/Si matrix. An environmental barrier coating (EBC) consisting of a silicon bond coat, a mullite intermediate coat, and a proprietary rare earth silicate topcoat was deposited on the SiC/SiC guide vane. Cooling holes were machined in the trailing edge using laser drilling prior to the application of the EBC. Three guide vanes with Haynes 188 superalloy vanes were tested using NASA's High-Pressure Burner Rig (HPBR) for 50 hrs of steady-state operation and 102 thermal cycles. The minimum gas temperature ranged from 900 to 1050°C, and the maximum temperature ranged from 940 to 1440°C. After 50 hrs testing, about 26% of the total coating surface area had spalled. After thermal cycle tests, melting of the superalloy vanes occurred; however, the SiC/SiC vanes were intact with deposits on the EBC surface.

In China, the National Key Lab of Thermostructure Composite Materials at Northwestern Polytechnical University designed and tested the CVI SiC/SiC high-pressure turbine twin-guide vanes [18], as shown in **Figure 10**. The 2D lamination with Z-direction puncture fiber preform was adopted for the guide vane with the fiber volume of 40–45% in total. The PyC interphase was deposited on the surface of the SiC fiber with a thickness of 300–500 μm. After the deposition of the SiC matrix using the CVI method, the SiC coating was deposited on the guide vane using the chemical vapor deposition (CVD) method to protect the internal fibers from oxidation under thermal shock tests. Cyclic thermal shock tests at target temperatures of 1400, 1450, and 1480°C in gas environment were conducted to analyze the damage mechanisms and failure

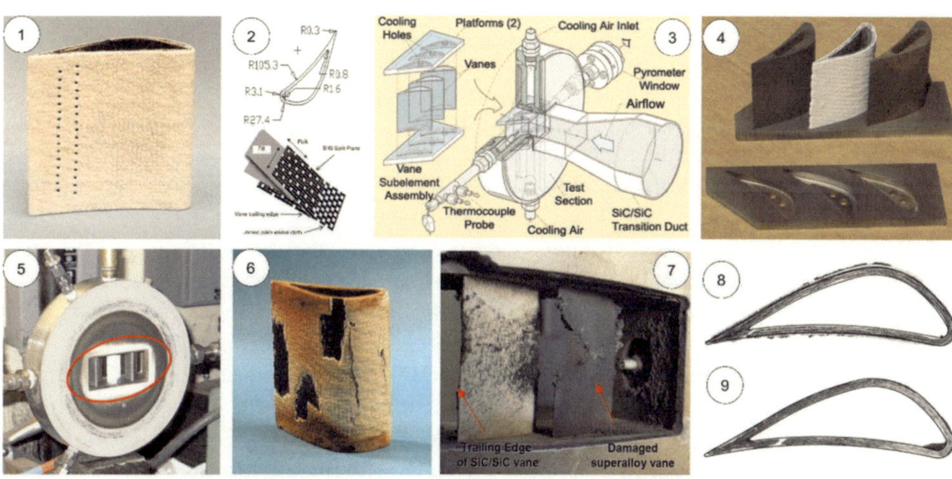


**Figure 9.**
*NASA GRC development and testing of CVI/slurry-cast/melt-infiltrated SiC/SiC turbine guide vanes. ① SiC/SiC guide vane with EBC and cooling holes. ② Geometry of the guide vane and Y-cloth architecture. ③ Configuration of testing section. ④ SiC/SiC guide vane and superalloy vanes. ⑤ Installed SiC/SiC vanes and superalloy vanes in the test section. ⑥ SiC/SiC guide vane after steady state rig testing underwater exposure for 3.5 h. ⑦ SiC/SiC guide vane and superalloy vanes after thermal cycle tests. ⑧ NDE inspection of SiC/SiC vane after thermal cycle tests. ⑨ NDE inspection of SiC/SiC vane after steady state rig testing.*

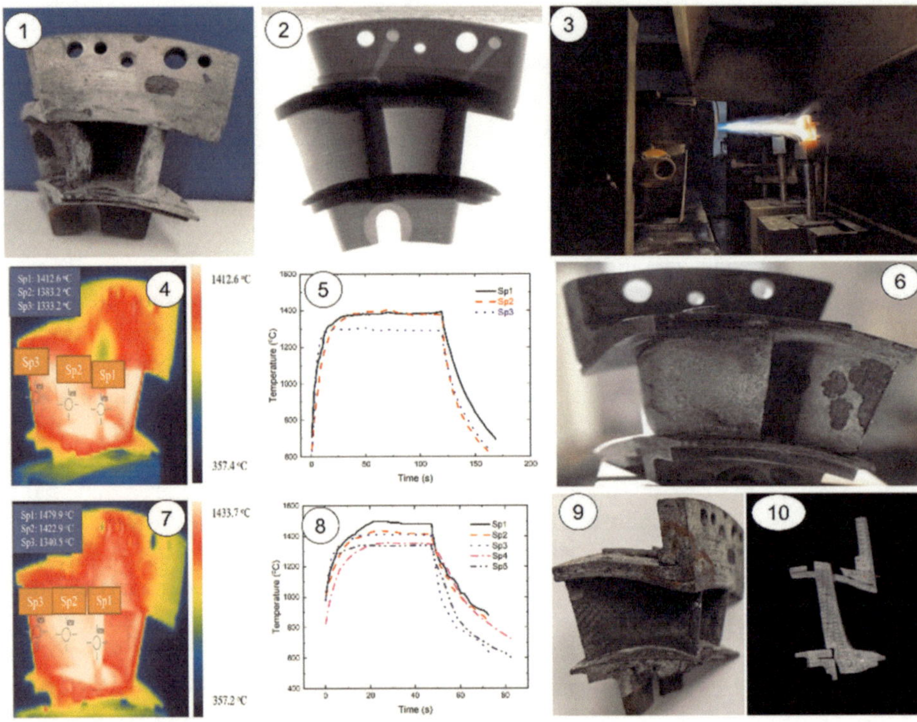


**Figure 10.**
*Design, fabrication, and testing of CVI SiC/SiC high-pressure turbine twin-guide vane by the National key lab of Thermostructure composite materials in Northwestern Polytechnical University [18]. ① SiC/SiC turbine guide vane after fabrication and machining. ② NDE observation of SiC/SiC turbine guide vane before the testing. ③ Thermal shock testing of SiC/SiC turbine guide vane. ④ Surface temperature distribution at a target temperature of 1400°C. ⑤ Monitored temperature changes during a thermal shock cycle for 1400°C. ⑥ Surface observation after 400 thermal shock cycles at 1400°C. ⑦ Surface temperature distribution at target temperature of 1480°C. ⑧ Monitored temperature changing during a thermal shock cycle for 1480°C. ⑨ Surface observation after 200 thermal shock cycles at 1480°C. ⑩ NDE observation of SiC/SiC turbine guide vane after 200 thermal shock cycles at 1480°C.*

modes. When the temperature increased from 1400 to 1480°C, the spalling of the SiC coating occurred at the guide vane, and delamination occurred at the area near the trailing edge of the guide vane. Under the observation of X-ray computed tomography (XCT), no obvious damage inside of the guide vane was observed.

In Japan, the IHI Corporation developed the SiC/SiC turbine guide vane and tested it in the IHI IM270 industrial gas turbine engine [19], as shown in **Figure 11**. The braided fabrics were used for the hollow airfoil portion, and 2D-woven cloth and 3D-orthogonal woven fabrics were used for the outer/inner platform portion. The Tyranno ZMI SiC fiber with the boron nitride (BN) interphase reinforced SiC matrix was fabricated using the chemical vapor infiltration (CVI), solid phase infiltration (SPI), and polymer impregnation and pyrolysis (PIP) methods. Cyclic thermal shock test for SiC/SiC vane was conducted by burner rig. The maximum temperature was set to 1200°C. One thousand thermal cycles were conducted with 6 mins for each thermal cycle. After thermal cyclic tests, the cracks were found at the leading edge due to the machining. The engine test was conducted on the IHI IM270 with four SiC/SiC vanes installed into the first-stage turbine nozzle assembly. The engine test was conducted for 413 hrs with 48 starts and normal stops. The inlet gas temperature was about 1050° C at steady state.

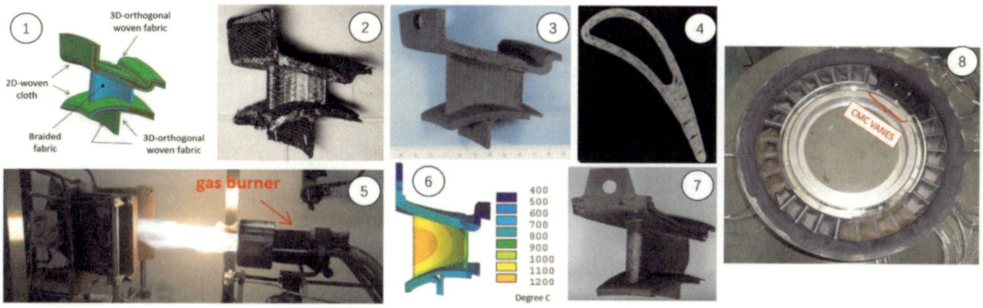


**Figure 11.**
*Design, fabrication, and testing of SiC/SiC turbine guide vane for the first stage of the IHI-IM270 engine [19]. ① Fiber arrangement of the SiC/SiC turbine guide van. ② SiC/SiC turbine guide vane as molded. ③ SiC/SiC turbine guide vane after machining. ④ X-ray CT of SiC/SiC turbine guide vane. ⑤ Burner rig test of SiC/SiC turbine guide vane. ⑥ Temperature distribution during burner rig test. ⑦ SiC/SiC turbine guide vane after burner rig test. ⑧ Engine test on IHI IM270.*

**Figure 12.**
*Cerasep® A40C low-pressure turbine blade developed by Herakles [20].*

### 2.2.2 Turbine blades

In France, Herakles has designed and manufactured SiC/SiC (Cerasep® A40C) turbine blades for the low-pressure turbine of a CFM56-5B engine [20], as shown in **Figure 12**. The subsequent tests of SiC/SiC blades performed by SNECMA in 2010 in an engine were a world first.

In Japan, the IHI Corporation designed and tested the SiC/SiC low-pressure turbine blade [21], as shown in **Figure 13**. The Tyranno ZMI SiC fibers were selected as the reinforcing fibers, and the boron nitride (BN) interphase was deposited on the SiC fiber by the CVI method to protect the SiC fibers from oxidation at elevated temperature. The fiber architecture for the main airfoil portion of the turbine blade was 3D orthogonal woven fabric with a fiber volume of 44%. The SiC matrix was fabricated using the chemical vapor infiltration (CVI), solid phase infiltration (SPI), and polymer impregnation and pyrolysis (PIP) methods. Three different types of dovetails were designed and tested under static pull at room temperature.

Type A dovetail with 3D orthogonal fabric were fractured in transverse shear mode at dovetail tongue, and the fracture loads for two specimens were 19.3 and 27.3 kN. Type B dovetail with 3D orthogonal fabric and 2D fabric was fractured in tensile mode at dovetail neck, and the fracture strength was the highest, i.e., 36.3 kN. Type C dovetail with 3D fabric was fractured in the transverse shear mode with a fracture strength of 16.7 kN. Cyclic load tests for type A, B, and C dovetail with the same

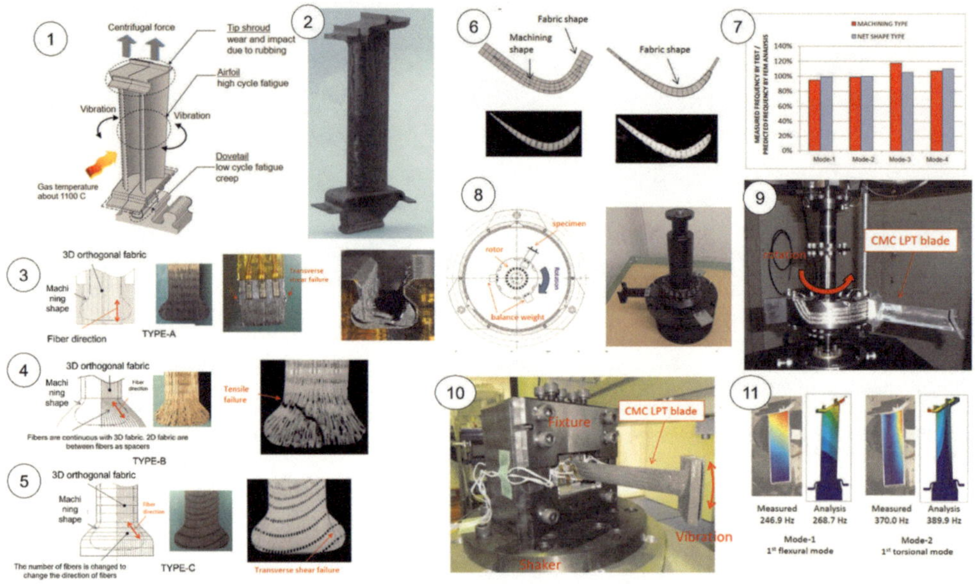


**Figure 13.**
*Design, fabrication, and testing of SiC/SiC low-pressure turbine blade by IHI Corporation [21]. ① Configuration and load types of low-pressure turbine blade. ② Prototype of SiC/SiC turbine blade. ③ Type A dovetail and failure mode under static pull test. ④ Type B dovetail and failure mode under static pull test. ⑤ Type C dovetail and failure mode under static pull test. ⑥ Machining type and net shape type for the airfoil portion. ⑦ Natural frequency of machining type and net shape type for different modes. ⑧ Spin test for tip shroud. ⑨ Spin test for turbine blade. ⑩ High cycle fatigue test for turbine blade. 11 Measured mode shapes and natural frequencies with FEM analysis.*

configuration were conducted under the peak load of 4.7 kN, minimum load of 0.49 kN, and a loading frequency of 0.5 Hz at room temperature. No defect was found after 15,000 cycles for the three types of dovetail. After the static pull tensile and cyclic load tests, Type B dovetail was chosen for the creep and cyclic test under pull load of 4.7 kN at elevated temperature of 650°C. Under cyclic load, the Type B dovetail specimen experienced 100,000 cycles without failure, and under creep load, Type B dovetail specimen tested for 500 hrs without failure. The durability of type B dovetail was proved.

For the airfoil portion, the machining type and net shape type were manufactured. Modal tests were conducted for the two airfoil types, and the natural frequency of Mode 1 and 2 were measured and compared with the finite element method (FEM) analysis.

For the tip shroud, type A fins hooked onto folded fabric and type B fins stitched on the middle position of the shroud were designed and manufactured. Spin tests and rubbing tests were conducted for the type A and B tip shrouds at room temperature. During the spin test, both type A and B specimens remained stable for 5 mins under a centrifugal load at 130% maximum rotation speed in an actual engine. During the rubbing tests, the maximum amount of wear was about 0.5 mm.

For the SiC/SiC low-pressure turbine blade, the spin test was conducted to check the static strength. No defect was found after the steady state for 5 mins under the centrifugal load at 120% maximum rotation speed in the actual engine. A high cycle fatigue test was conducted at the first flexural mode frequency. After $10^7$ cycles, no defect was found, and the natural frequency did not change.

### *2.2.3 Turbine blisk*

In the United States, under the NASA SIMPLEX Turbopump Blisk program, the C/SiC blisk prototype for rocket engine was manufactured by NASA Glenn Research Center (GRC) and George C. Marshall Space Flight Center (MSFC) using the CVI method and tested in the SIMPLEX Turbopump at NASA-MSFC Test Stand [22], as shown in **Figure 14.** One blisk was constructed with a polar woven fiber preform, and

**Figure 14.**
*C/SiC turbine blisk for rocket engine developed by NSAS-GRC [22]. ① C/SiC turbine blisk installed on a shaft. ② SIMPLEX Turbopump. ③ C/SiC turbine blisk. ④ CT image of polar woven C/SiC blisk.*

**Figure 15.**
*SiC/SiC turbine blisk developed by Japan [23]. ① SiC/SiC turbine blisk. ② Rig test of SiC/SiC turbine blisk.*

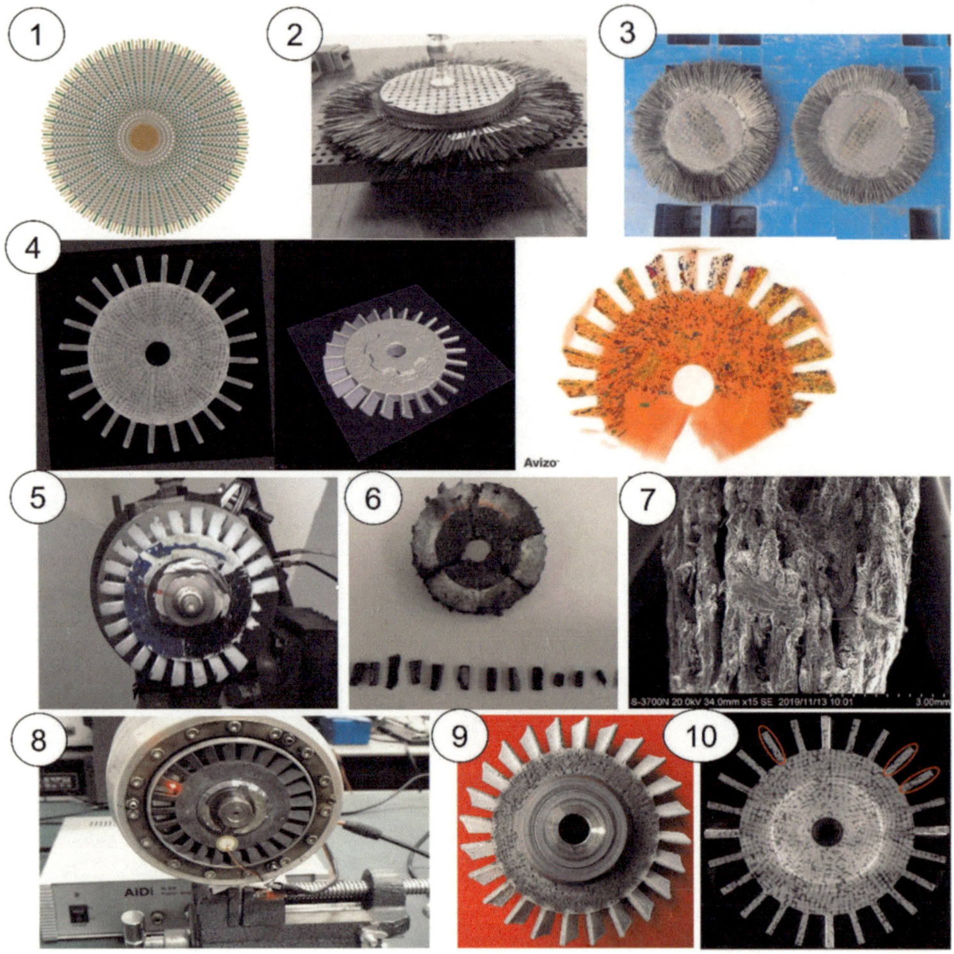


**Figure 16.**
*Design, fabrication and testing of SiC/SiC turbine blisk [24]. ① Schematic of the spider web structure (SWS) fiber preform for turbine blisk. ② SWS fiber preform for turbine blisk. ③ SWS fiber preform with deposited SiC matrix using CVI method. ④ CT detection and 3D reconstruction analysis of SiC/SiC turbine blisk. ⑤ Overspeed rotation test of SiC/SiC turbine blisk at room temperature. ⑥ Blade flying off of the SiC/SiC turbine blisk. ⑦ Fragment scanning electron microscope (SEM) image. ⑧ SiC/SiC turbine blisk installation for the engine test. ⑨ Surface observation of SiC/SiC turbine blisk after engine test. ⑩ CT detection of SiC/SiC turbine blisk after engine test.*

the second blisk was constructed with a quasi-isotropic preform. Two C/SiC blisks were tested in a rig under environmental conditions. The CT scan was used to analyze the internal defects of the fabricated blisk. The 2D and 3D C/SiC bench-top test specimens and C/SiC blisk specimens with quasi-isotropic and polar woven fiber architectures showed that the material was very lightly damped with high frequency-dependent damping.

In Japan, under the research and development program of Advanced Material Gas-Generator (AMG) initiated in 1993, the SiC/SiC turbine blisk was machined from the disk fabricated by 3D woven SiC fiber densified with SiC matrix using CVI and PIP method [23], as shown in **Figure 15**. The blisk was mounted to the metallic shaft and rotated at 30000 RPM (tip rotational speed of 386 m/s) by exposing blade section to 973 K gas stream with no vibration and no damage observed.

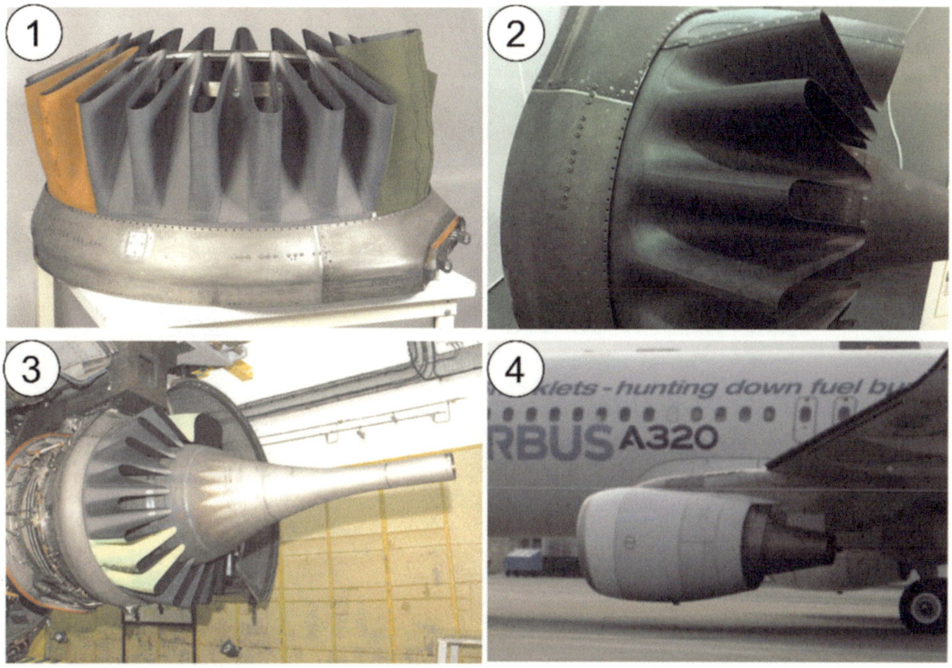


**Figure 17.**
*Cerasep® A40C mixer developed by SAFRAN [20]. ① Cerasep® A40C mixer demonstration. ② Side photo of Cerasep® A40C mixer demonstration. ③ Cerasep® A40C mixer installed on CFM56-5C engine. ④ Flight testing of Cerasep® A40C mixter on A320.*

In China, the National Key Lab of Thermostructure Composite Materials in Northwestern Polytechnical University designed and tested the SiC/SiC turbine blisk [24], as shown in **Figure 16**. During the engine bench test, low cyclic fatigue of $N$ = 994 cyclic with maximum speed $n_{max}$ = 60,000 RPM and $N$ = 100 cyclic with maximum speed $n_{max}$ = 70,000 RPM were completed.

### 2.3 Application of exhaust components

In France, the first technology demonstration dedicated to the evaluation of a CMC mixer for a CFM56-5C engine, powering the A340–200/300 jets, was conducted by SAFRAN [20], as shown in **Figure 17**. The component showed a 30% weight saving over Inconel. The prototype mixer manufactured with Cerasep® A40C was ground tested in 2007 and completed 700 engine cycles and 70 take-off hours with no material damage identified.

In the United States, the NASA Glenn Research Center (GRC) and Rolls-Royce Liberty Works (RRLW) teamed with ATK-COIC in the NASA ERA (Environmentally Responsible Aviation) project with the goal of advancing oxide/oxide mixer nozzle technology to full-scale engine testing [25], as shown in **Figure 18**. A subscale oxide/oxide exhaust mixer with 10 lobes and a diameter of approximately 200 mm was manufactured and tested in several rigs. The vibration test was completed in the GRC Structural Dynamic Lab in July 2014, including sine sweeps and dwells. Two 1 million cycle dwells were performed at room temperature, and one 100,000 cycle dwell was performed at 371°C. The deformation of the lobes was observed, and only three minor defects were observed.

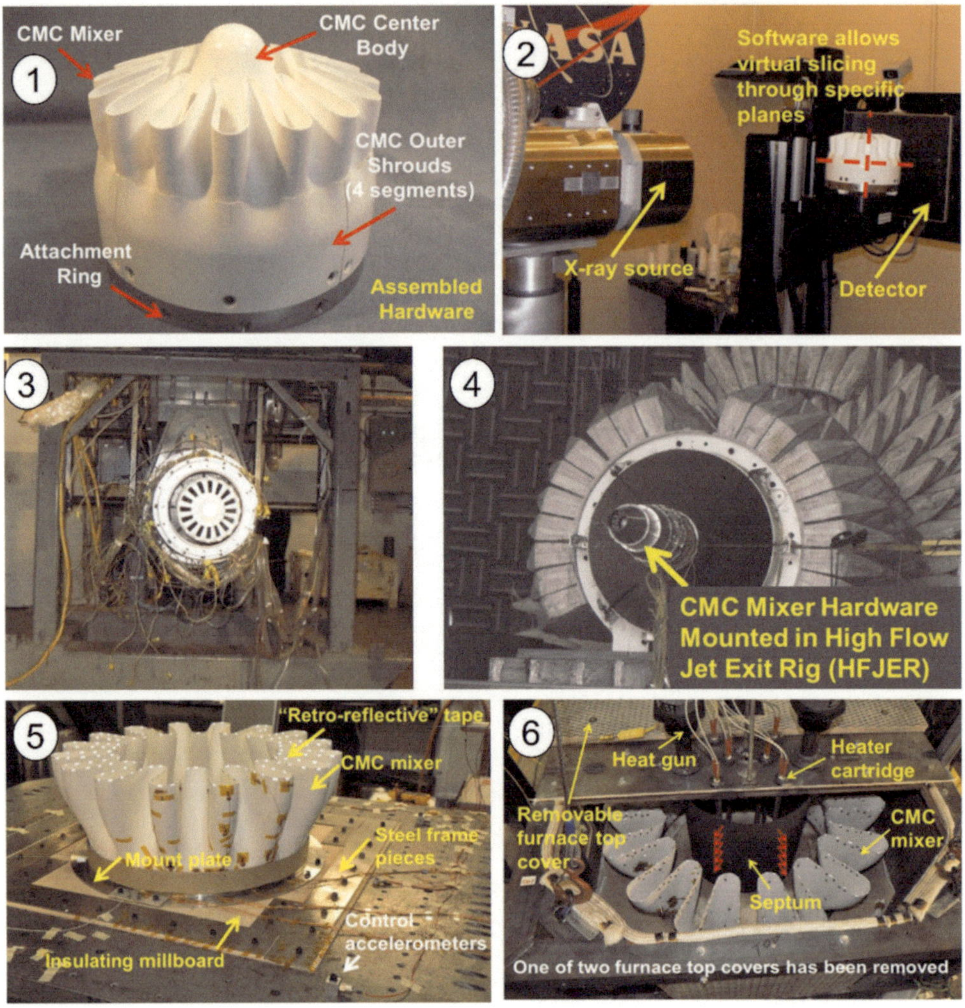


**Figure 18.**
*Oxide/oxide exhaust mixer developed and tested under the NASA ERA program [25]. ① Oxide/oxide subscale mixer and center body. ② NDE for the subscale oxide/oxide mixer and outer shrouds. ③ Performance testing of oxide/oxide exhaust mixer. ④ Nozzle acoustic test rig with oxide/oxide exhaust mounted. ⑤ Vibration testing of full-scale oxide/oxide exhaust mixer at room temperature. ⑥ Vibration testing of full-scale oxide/oxide exhaust mixer at elevated temperature.*

Under the continuous Lower Energy, Emissions, and Noise (CLEEN) Technologies Development Program, the Boeing Company is working to produce an acoustic Nextel 610/aluminosilicate composite centerbody and exhaust nozzle for commercial aircraft [26, 27], as shown in **Figure 19**. In 2014, Boeing tested the oxide/oxide nozzle on a 787 aircraft. This nozzle technology can withstand higher temperatures, and lower fuel consumption, and can also accommodate acoustic treatments that reduce community noise.

GE company developed an oxide/oxide exhaust mixer of 610 mm long and 965 mm in diameter and core cowling for Passport 20 engine powered by the Bombardier Global 7000 and 8000 ultra-long-range twinjets [28], as shown in **Figure 20**. The oxide/oxide exhaust mixer enabled a weight saving of approximately 20 ks compared to the metallic baseline.

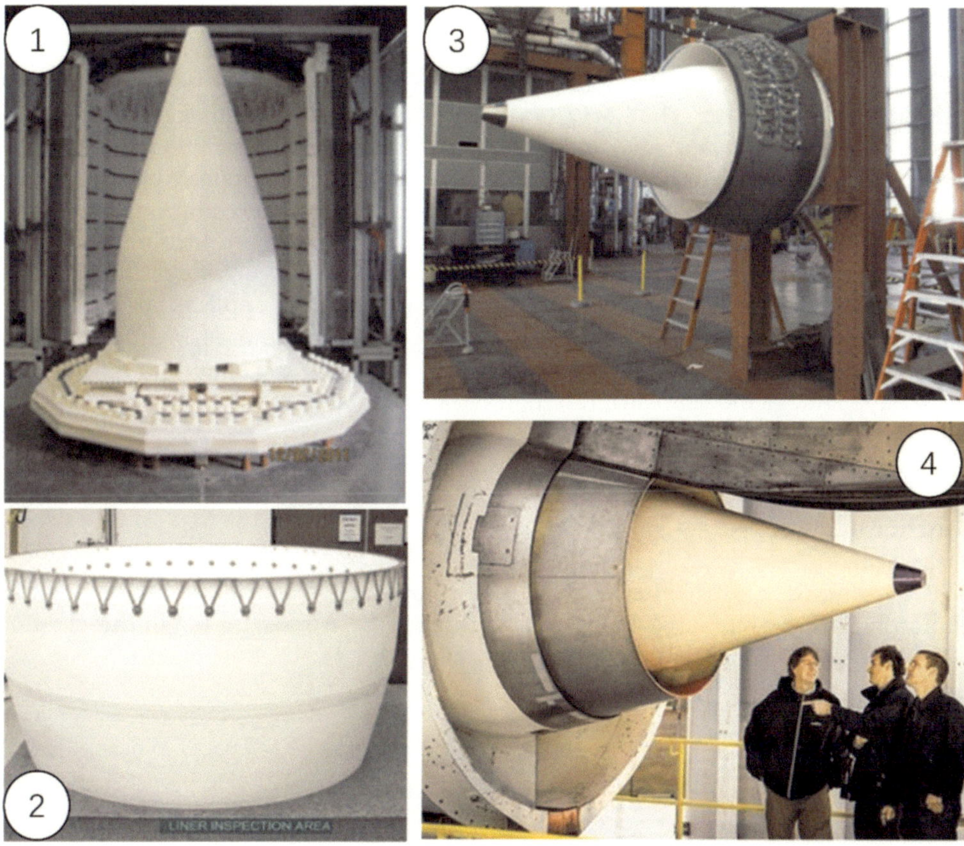


**Figure 19.**
*Development and testing of oxide/oxide exhaust nozzle for the Trent 1000 engine by Boeing Company under CLEEN program of the FAA [23]. ① Oxide/oxide centerbody. ② Oxide/oxide outer ring. ③ Oxide/oxide exhaust nozzle assembly. ④ Ground engine testing of the oxide/oxide exhaust on Trent 1000.*

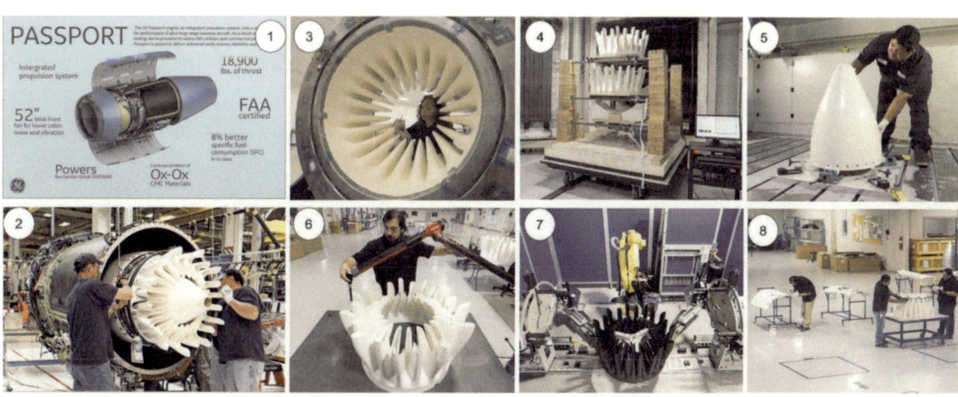


**Figure 20.**
*Development, fabrication, and installation of an oxide/oxide exhaust mixer for Passport 20 engine [28]. ① Passport 20 engine developed by GE company. ② Installation of oxide/oxide exhaust mixer on Passport 20 engine. ③ Oxide/oxide hand layup on metal molds. ④ Oxide/oxide mixer after curing at sintering furnace. ⑤ Milling, grinding and drilling of oxide/oxide centerbody. ⑥ Quality control checking on the outer surface of the mixer. ⑦ Nondestructive inspection on oxide/oxide mixer using infrared ash thermography. ⑧ Assembly of the metal and ceramic parts.*

## 3. Summary and conclusion

Ceramic-matrix composites possess high specific strength and modulus, especially at elevated temperatures. The CMCs hot-section components were developed by France, USA, China, Japan, etc., and have already been applied in military or commercial aero engines. This chapter reviews the recent design, application, and challenges of ceramic-matrix composites in hot-section components for gas turbine engines, including the CMC combustor liner, turbine guide vanes, turbine blades, turbine blisk, and exhaust mixer. In these applications, the non-oxide SiC/SiC and oxide/oxide CMCs were the main composites for engineering applications in hot-section components of aero engines. In the future, more and more CMC components will be used in commercial and military engines. To ensure the operation reliability and safety, damage mechanisms, failure modes and related models and prediction tools should be developed.

## Conflict of interest

The author declares that he has no known competing financial interests or personal relationships that could have appeared to influence the work reported in this chapter.

## Author details

Longbiao Li
College of Civil Aviation, Nanjing University of Aeronautics and Astronautics, Nanjing, P.R. China

*Address all correspondence to: llb451@nuaa.edu.cn

## References

[1] Li LB. Ceramic Matrix Composites: Lifetime and Strength Prediction under Static and Stochastic Loading. Oxford, UK: Elsevier; 2023

[2] Li LB. Nonlinear Behavior of Ceramic-Matrix Composites. Oxford, UK: Elsevier; 2021

[3] Guo X, Wu J, Li J, Zeng Y, Huang X, Li LB. Damage monitoring of 2D SiC/SiC composites under monotonic and cyclic loading/unloading using acoustic emission and natural frequency. Ceramics-Silikaty. 2021;**65**:125-131

[4] Liu H, Li L, Wang Y, Zhou Y, Ai Y, Yang J, et al. In situ damage propagation and fracture in notched cross-ply SiC/SiC composites: Experiment and numerical modeling. Journal of the European Ceramic Society. 2024;**44**:2052-2064

[5] Li L. Characterization of multi-step cyclic-fatigue hysteresis behavior of 2D SiC/SiC composite using inverse tangent modulus. Ceramics International. 2024;**50**:4392-4403

[6] Lü X, Li L, Sun J, Yang J, Jiao J. Microstructure and tensile behavior of $(BN/SiC)_n$ coated SiC fibers and SiC/SiC minicomposites. Journal of the European Ceramic Society. 2023;**43**:1828-1842

[7] Lacombe A, Spriet P, Alain A, Bouillon E, Habarou G. Ceramic matrix composites to make breakthroughs in aircraft engine performance. In: 50th AIAA/ASME/ASCE/AHS Structural Dynamic, and Materials Conference; May 4–7, 2009; Palm Springs, California, USA. Reston, VA, USA: The American Institute of Aeronautics and Astronautics; 2009

[8] Misra AK. Development of advanced engine materials in NASA ultra efficient engine technology program. In: Louisiana:15th International Symposium on Air Breathing Engines Conference. 2001

[9] Padture NP. Advanced structural ceramics in aerospace propulsion. Nature Materials. 2016;**15**:804-809

[10] Bhatia T, Jarmon D, Shi J, et al. CMC combustor liner demonstration in a small helicopter engine. In: ASME Turbo Expo 2010: Power for Land, Sea, and Air. Glasgow: American Society of Mechanical Engineers Digital Collection; 2010. pp. 509-513

[11] Van Roode M, Price J, Kimmel J, Miriyala N, Leroux D, Fahme A, et al. Ceramic matrix composite combustor liners: A summary of field evaluations. Journal of Engineering Gas Turbines Power. 2007;**129**:21-30

[12] Suzuki Y, Satoh T, Kawano M, Akikawa N, Matsuda Y. Combustion test results of an uncooled combustor with ceramic matrix composite liner. In: ASME Turbo Expo 2001: Power for Land, Sea, and Air; June 4–7, 2001; New Orleans, Louisiana, USA. New York, NY, USA: The American Society of Mechanical Engineers; 2001

[13] Matsuda Y, Akikawa N, Satoh T. Manufacturing of 3-D woven $SiC_{f/}SiC$ composite combustor liner. In: 25th Annual Conference on Composites, Advanced Ceramics, Materials, and Structures: A: Ceramic Engineering and Science Proceedings. Hoboken, New Jersey, USA: John Wiley & Sons, Ltd; 2001. pp. 463-470

[14] Gerenda M, Cadoret Y, Wilhelmi C, et al. Improvement of oxide/oxide CMC and development of combustor and

turbine components in the HiPOC program. In: Proceedings of ASME Turbo Expo 2011; June 6–10, 2011; Vancouver, British Columbia, Canada. New York, NY, USA: The American Society of Mechanical Engineers; 2011

[15] Gerendas M, Wilhelmi C, Machry T, et al. Development and validation of oxide/oxide CMC combustors within the HiPOC program. In: Proceedings of ASME Turbo Expo 2013: Turbine Technical Conference and Exposition; June 3–7, 2013; San Antonio, Texas, USA. New York, NY, USA: The American Society of Mechanical Engineers; 2013

[16] Behrendt T, Hackemann S, Mechnich P, et al. Development and test of oxide/oxide ceramic matrix composites combustor liner demonstrators for aero-engines. Journal of Engineering for Gas Turbines and Power. 2017;**139**(3):031507

[17] Verrilli M, Calomino A, Craig RR. Ceramic matrix composite vane subelement testing in a gas turbine environment. In: Proceedings of ASME Turbo Expo 2004, Power for Land, Sea, and Air; June 14–17, 2004; Vienna, Austria. New York, NY, USA: The American Society of Mechanical Engineers; 2004

[18] Liu X, Guo X, Xu Y, Li LB, Zhu W, et al. Cyclic thermal shock damage behavior in CVI SiC/SiC high-pressure turbine twin guide vanes. Materials. 2021;**14**:6104

[19] Fumiaki W, Takeshi N, Yousuke M. Design and testing for ceramic matrix composite turbine vane. In: Proceedings of ASME Turbo Expo 2017: Turbomachinery Technical Conference and Exposition GT2017; June 26–30, 2017; Charlotte, NC, USA. New York, NY, USA: The American Society of Mechanical Engineers; 2017

[20] Patrick S. CMC applications to gas turbines. In Bansal NP, Lamon J: Ceramic Matrix Composites – Materials, Modeling and Technology. 1st ed. Hoboken, New Jersey: John Wiley & Sons, Inc.; 2014

[21] Watanabe F, Nakamura T, Shinohara K. The application of ceramic matrix composite to low pressure turbine blade. In: Proceedings of ASME Turbo Expo 2016: Turbomachinery Technical Conference and Exposition, GT2016; June 13–17, 2016; Seoul, South Korea. New York, NY, USA: The American Society of Mechanical Engineers; 2016

[22] Min J, Harris DL, Ting JM. Advances in ceramic matrix composite blade damping characteristics for aerospace turbomachinery applications. In: 52nd AIAA/ASME/ASCE/AHS/ASC Structures, Structural Dynamics and Materials Conference; 19th April 4–7, 2011; Denver, Colorado. Reston, VA, USA: The American Institute of Aeronautics and Astronautics; 2011

[23] Kuriyama T, Miyagawa H, Uyama M, Yamamoto S, Yokoi S, Hiromatsu M. Status of AMG (Advanced Material Gas-Generator) research and development program. In: Proceedings of ASME TURBO EXPO 2001; June 4–7, 2001; New Orleans, Louisiana, USA. New York, NY, USA: The American Society of Mechanical Engineers; 2001

[24] Liu X, Xu Y, Li J, et al. Design, fabrication and testing of ceramic-matrix composite turbine blisk. Acta Materiae Compositae Sinica. 2023;**40**(3): 1696-1706

[25] Kiser JD et al. Oxide/oxide ceramic matrix composite (CMC) exhaust mixer

development in the NASA environmentally responsible aviation (ERA) project. In: ASME Turbo Expo 2015: Turbine Technical Conference and Exposition, Montreal, Quebec. New York, NY, USA: The American Society of Mechanical Engineers; 2015

[26] Boeing. Continuous lower energy, emissions and noise (CLEEN) technologies development. In: The CLEEN Consortium Meeting. Arlington County, Virginia, USA: Boeing; 2011

[27] Petervary M, Steyer T. Ceramic matrix composites for structural aerospace applications. In: 4th International Congress on Ceramics. Chicago, IL; 2012

[28] Composites World: Ceramic Matrix Composites: Hot Engine Solution. 2017. Available from: https://www.compositesworld.com/articles/ceramic-matrix-composites-hot-enginesolution [Accessed: December 21, 2021]

Chapter 2

# Application of Acoustic Emission to Detect Damage in Composites Materials

*Sattar Mohammadi Esfarjani*

## Abstract

In today's modern world, the use of composite in the construction of various equipment and parts due to many advantages such as; weight reduction, design durability, flexibility, etc., is increasing. Damages may occur unintentionally in composite materials, either during the manufacturing process or during the normal lifetime of the part. Structural health monitoring (SHM) of composite structures is an industry requirement. The acoustic emission method can be used as an effective nondestructive testing (NDT) method to continuously monitor the development of damages in composites. In this chapter of the book, it has been prepared with the aim of application of AE to detect damage in composites. For this purpose, the literature done in the field of SHM of composite structures using the AE method is reviewed. The content of this chapter shows the importance of using the AE method for SHM of composite structures.

**Keywords:** structural health monitoring, damage detection, nondestructive testing, acoustic emission, composite

## 1. Introduction

Damage can occur in parts during the manufacturing process or after completion of construction for various reasons such as; transportation, maintenance conditions, working conditions, etc. Therefore, it is necessary to ensure the health of the structure and identify the damage during the process of its construction and operation. Parts inspection methods are divided into destructive testing (DT) and nondestructive testing (NDT) categories. Compared to DT methods, NDT is a diagnostic method without introducing damage, stress or failure into the test. Because NDT methods do not damage the part, it saves time and cost for testing [1–4].

Nowadays, NDT methods are very important in the industry. Many standards and technical instructions use NDT methods for a wide range of equipment and devices such as; storage tanks, pressure tanks, pressure pipes, metal structures of tall buildings, bridges, etc., have been considered mandatory [5–8].

By using NDT techniques, a cost-effective tool can be accessed to test a sample for individual investigation and examination. Sometimes it takes more than one NDT test method to ensure the structural health monitoring (SHM) of a part. Therefore,

| Method | Advantages | Disadvantages |
|---|---|---|
| VT | • Cheap<br>• Easy to train<br>• Portable<br>• Minimum part preparation | • Surface indications only<br>• Generally, only able to detect large flaws<br>• Possible misinterpretation of flaws |
| PT | • Wide application range for different genders and shapes<br>• Very low cost of implementation<br>• Completely portable<br>• Direct appearance of works on the surface<br>• Simple application<br>• Regardless of the shape and dimensions of the piece<br>• Full testing of the part is possible<br>• It requires relatively little operator training.<br>• Very sensitive to surface damage<br>• Low investment cost<br>• Can be automated | • Only the ability to identify surface defects<br>• Not to be used on porous, sanded and machined surfaces<br>• Application limitation for hot objects<br>• The scope of work should be available<br>• The galvanized and colored layers on the surface of the part must be removed.<br>• Cleaning work and surface operations are necessary<br>• Determining the depth of damage is rarely possible<br>• Larger cracks are not visible |
| MT | • Low cost of implementation<br>• The ability to identify subsurface defects<br>• Direct appearance of works on the surface<br>• The ability to perform tests despite the color coating<br>• The method is fast and reliable<br>• It is possible to identify the most subtle surface defects<br>• It is almost independent of the shape and dimensions of the piece<br>• It requires relatively little operator training.<br>• The defect is directly visible | • Application only for ferromagnetic materials<br>• The need for several inspection steps in different directions<br>• A demagnetization step is required<br>• High current requirement<br>• Only defects close to the surface can be proven<br>• It is not possible to assess the depth of defects<br>• Thick surface coatings should usually be removed<br>• The test range should be available |
| RT | • Application for all materials permanent test record obtained not only for surface methods but also covered parts can be tested<br>• Transcription of the findings in the form of a film<br>• The size and shape of the damage can be seen<br>• Direct access to the test area is not necessary<br>• Especially suitable for three-dimensional damages such as; holes, heterogeneities, heat cracks and welding defects | • Decrease in sensitivity with increasing thickness, high degree of skill and experience required for exposure and interpretation need full access to the part<br>• Only limited thicknesses can be tested<br>• The ability to detect damage is relative to the thickness of the radiation.<br>• Sensitive to flaw orientation<br>• Determining the depth of damage is difficult.<br>• It is a costly and time-consuming method<br>• There is a risk of radiation |

| Method | Advantages | Disadvantages |
|---|---|---|
| UT | • Immediate results<br>• Identification of surface and subsurface defects<br>• Minimal surface preparation<br>• Determining the depth of the crack<br>• Wide range of materials and thickness can be tested.<br>• Commenting on the shape, type and size of the damage is possible<br>• Fully automatic testing is possible<br>• Sensitive to very small discontinuities | • High degree of skill required to set up and interpret<br>• It is difficult to visit uneven, irregularly shaped, very small and very thin materials<br>• The need for reference standards to calibrate equipment and crack characteristics<br>• Weakness in identifying linear defects parallel to the sound beam<br>• Surface must be accessible to probe<br>• Couplant usually required<br>• It is used only purposefully and for certain damages and a certain range<br>• The test range must be available (audio connection). |
| TT | • It is fast and safe.<br>• Can identify of a defect in the part before it completely fails.<br>• Large areas can be scanned fast<br>• It does not need to contact the surface of the piece. | • It only shows surface defects<br>• Need an experienced expert<br>• The results are not very accurate. |
| Leak testing (LT) | • They are cheaper than some non-destructive testing methods such as; RT and UT.<br>• It is safe. | • It only shows surface defects<br>• Need an experienced expert<br>• The results are not very accurate. |
| Guided wave testing (GWT) | • Discovery of surface and subsurface defects in the part<br>• The test data is completely recorded.<br>• It is fast<br>• It is portable.<br>• Minimal surface preparation required | • We need an experienced expert.<br>• Difficult to find small pitting defects. |
| Eddy current testing (ET) | • Fast method<br>• Sensitivity to surface defect<br>• Can detect through surface coatings<br>• Little pre-cleaning required<br>• The penetration depth is a function of variable frequency<br>• Fully automatic testing and fully automatic evaluation are possible<br>• Portability | • Only electrically conductive materials can be tested<br>• Very small damages are difficult to identify<br>• Large-scale testing is time-consuming<br>• Detection is affected by changes in shape and conductivity<br>• The damage is not directly visible<br>• It needs a lot of experience<br>• Complex designs are not inspected with accurately.<br>• Heat loss is more.<br>• No permanent record (unless automated) |

**Table 1.**
*Advantages and disadvantages of NDT methods [9, 12, 13].*

it is necessary to familiar with the performance, advantages and limitations of NDT methods to ensure the success of the evaluation [9].

There are different types of NDT methods. Each test method has its own scope of application. Therefore, according to the sample material and the type of expected damage, the appropriate method should be chosen. Often, using a combination of several methods has also been helpful. NDT methods also have limitations. For example, improper part design can make NDT methods unable to detect damage. However, due to their simple and cost-effective use of conventional NDT methods, they are often considered important materials testing tools [9–11].

Various methods are used to ensure the SHM of composite structures. Several NDT techniques are used for SHM of composite, including; radiographic testing (RT), visual testing (VT) or visual inspection (VI), ultrasonic testing (UT), thermographic testing (TT), infrared thermographic testing, acoustic emission (AE) testing, acoustic-ultrasonic, electromagnetic test, optical test, penetration testing (PT) or liquid penetrant testing (LPT) and magnetic particle testing (MT). In **Table 1**, the advantages and disadvantages of some commonly used NDT methods are listed.

## 2. Acoustic emission method

When a solid material is under stress, the defects in it cause the creation of high frequency sound waves. These waves are emitted in the material and they can be received by piezoelectric sensors. By analyzing these waves, it is possible to identify the type of defect, its location and severity. This process forms the basis of the AE method. Therefore, accurate and correct identification of the collected signals and their correct analysis are very useful in making important decisions regarding the continuation of a process or stopping it and taking the necessary measures.

AE is a passive method. In this method, first the elastic waves produced in the material are received. Then, they are analyzed in order to communicate between the received waves and the changes made on the source. According to the propagation of waves from the source to the surface of the material, they can be recorded by sensors and in this way information about the existence and location of the source of propagation of waves can be obtained. These waves can have frequencies up to several megahertz. In this method, frequencies usually are in the range of 150–300 kHz. The propagation of waves continues to the surface of the piece, that is, where the sensors are installed. Then, they are recorded by sensors and they are converted into electrical signals. The AE system processes these signals and turns them into information

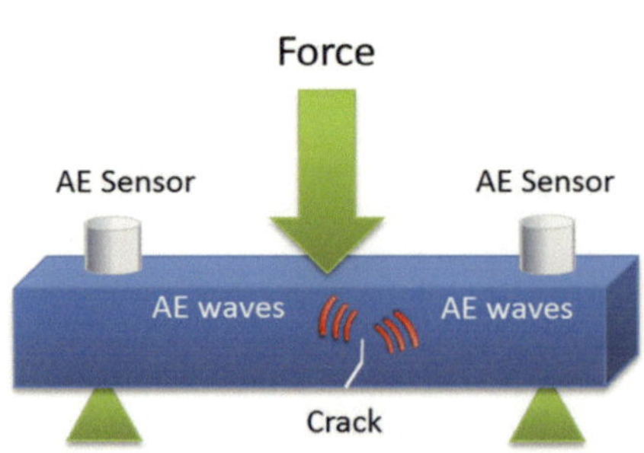


**Figure 1.**
*Creation of elastic waves due to stress in the material [14].*

| Advantages | Disadvantages |
|---|---|
| • It is able to detect very small defects in the range of 1 μm.<br>• Inspection with this method is very fast and as a result work efficiency is very high.<br>• This method is less sensitive to the part geometry.<br>• Compared to other conventional NDT testing methods, this method has a high investment value due to its higher efficiency.<br>• Can detect damages in defects that are difficult to access with conventional non-destructive testing techniques<br>• Deep effect<br>• Location of damage is possible<br>• It is simple to use<br>• Can be conducted remotely<br>• A non-invasive method | • In this method, static defects, defects that neither grow nor move can be recognized, although this limitation has been partially overcome by applying external stress.<br>• The damage should emit a sound<br>• The part must be loaded<br>• Sensitive to disturbing sounds<br>• Can be slower than other non-destructive testing techniques |

**Table 2.**
*The advantages and disadvantages of the AE method [31–33].*

packets. Finally, statistical information such as; the characteristics and location of resources are calculated and they are displayed in the form of graphic and numerical charts to be interpreted (**Figure 1**) [15, 16].

Among the applications of the AE method for SHM the following can be mentioned; detection of friction [17], detection of friction and the presence of wear in bearings (bearings, roller bearings) due to improper functioning of grease or lubricant, detection of impact in rotating mechanisms due to excessive looseness of the bearing that causes impact [18, 19], detection of turbulence in the inspection of tanks under it is the pressure that the presence of excessive leakage causes turbulent flow [20], detection of defects in manufactured parts [21, 22], detection of defects in welding [23] and estimation of the life of the structure [24, 25]. Also, this method can be used to reveal and locate partial voltage discharges in large transformers [26], research and investigate the properties and characteristics of materials [27], geology [28, 29] and research on micro-vibrations [30]. The advantages and disadvantages of the AE method are listed in the following **Table 2** [31–33].

## 3. Composite defects and their detection

In today's modern world, the use of composite in the construction of various equipment and parts due to many advantages such as; weight reduction, design durability and flexibility, etc., is increasing. The percentage of composite components in commercial aircrafts is shown in **Figure 2**. For example, In the 1970s, only 7% of Concorde airplanes were built with fiber-reinforced polymer (FRP) composites. Today, 80% of the Boeing 787 Dreamliner is made of FRP composites [35]. Composites are materials that are created from the combination of at least two materials with different physical and chemical properties. In this composition, one of

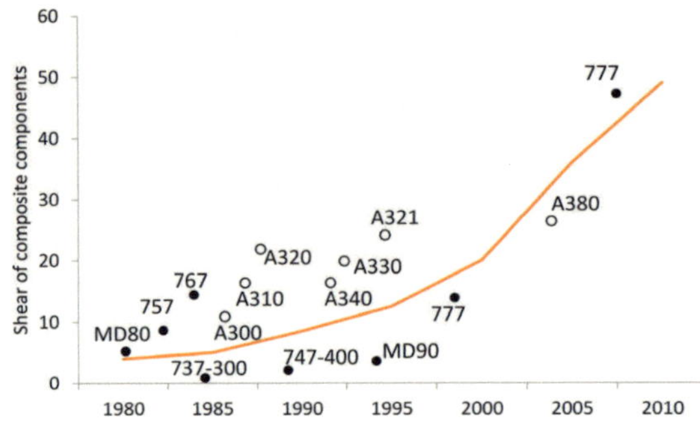


**Figure 2.**
*Percentage of composite components in commercial aircrafts [34].*

the components plays the role of background and the other plays the role of reinforcement. Composites are produced in different compositions. Composites are widely used in various industries. Composites have many applications in various industries such as; aerospace, construction, dentistry, electricity and power plants [3].

Damages may occur unintentionally in composite materials, either during the manufacturing process or during the normal lifetime of the part [36]. Porosity, foreign bodies, incorrect fiber volume fraction due to excess or insufficient resin, debonding defects, fiber misalignment, ply misalignment, incompletely cured matrix due to incorrect curing cycle or faulty material, wavy fibers, delamination, fiber defects, contamination, improper cure, resin rich/poor, voids, cracks, missing adhesive impact damage, thermal damage, thickness variation, dimensional problem and interface integrity are examples of damage in composites [9, 37].

## 4. SHM of composite structures using AE method

SHM of composite structures is an industry requirement. The AE method can be used as an effective NDT method to continuously monitor the development of fatigue damage in composites [38]. Fatigue damage caused by stress or thermal cycling can affect the mechanical integrity and safety of a composite structure. Therefore, it is necessary to identify the damage early using NDT methods [39]. Diagnosing damage in composites requires the improvement of old NDT methods as well as the development of new NDT methods. The AE method basically includes the detection of stress waves with low intensity that is caused by the occurrence of damage. For example, it can refer to warping and breakage of fibers. Stress waves are formed from the strain energy that is released during a failure and are transferred to the levels that are detected by acoustic sensors [40, 41]. One of the advantages of the AE method is the ability to monitor the onset and accumulated damage in real-time, which is not possible with most other NDT methods. Another advantage of the AE method is the ability to distinguish between some types of damage and, if a number of sensors are used, to determine the location of damage events. The AE method is often used to monitor the formation and growth of fatigue damage in composites [39, 42].

Previous studies have shown that inspection of composites using the AE method can provide valuable insight into failure mechanisms that occur at different stages during fatigue life and ultimately lead to failure. This is because different fatigue failure mechanisms create different AE signals. In composite materials, if changes occur due to loading, such as; delamination, separation of resin, breaking of reinforcing fibers, etc., AE signals are produced. In addition to the above, defects such as; corrosion and leakage of ultrasonic signals are also produced. For example, corrosion on the bottom of oil tanks produces explosive signals that propagate through the oil liquid to the wall of the tank. By installing sensors on the wall, these signals can be identified. The signals created by leakage can also be explosive and continuous. Basically, explosions occur at high pressure. When the pressure is low or the current is calm, continuous signals with low amplitude and small propagation distance are created. The ability of the AE test to identify these defects has caused to expand the use of this method in the inspection of equipment such as; composite pressure tanks, storage tanks and piping systems [43].

Godin et al. used the AE method as a method to detect different types of failures in glass-polyester composite materials. The main goal of their work was to analyze the signals in order to identify different sources of emitted waves [44, 45]. Özaslan et al. identified the damage mode of composite sheets based on the AE method and digital image correlation (DIC) method [46]. Muir et al. identified the damage mechanism in composites through machine learning and acoustic propagation [47]. Fotouhi et al. investigated the observation of the initiation and growth of interlayer separation failure in composite materials using the AE method [48]. Fotouhi et al. using the AE method and fuzzy classification and wavelet transform, classified the different failure mechanisms that occur during the growth of interlayer separation in glass/epoxy samples under three-point bending [49].

In the study of Yu et al. an AE signal analyzer with a resonant circuit was designed and built to extract the specified frequency of an AE signal. The results showed that the developed AE signal analyzer generally has the same crack detection capability as a conventional AE signal analyzer, under static and dynamic tensile tests of composite materials [50]. Chang et al. found that transverse matrix cracks caused by fatigue loading of carbon/epoxy composites can be identified by detecting high-amplitude sound emissions that have a specific frequency [51]. Barile et al. found that using the characteristics of the AE method, the failure and growth of interlayer cracks in high-amplitude signals can be detected using this test [52].

Hamzeloo et al. studied the degradation mechanisms in the bending of polymer-based composite sandwich panels by the AE method. By analyzing the acoustic data, they investigated and diagnosed the damage mechanisms in polyester/glass composite sandwich panels with polyurethane foam core in the three-point bending test by applying initial attenuation resulting from impacts with different energies [53].

The research of Saeedifar et al. investigated the creation and expansion of interlayer separation in glass/epoxy multilayer composites by the AE method. The results show the optimal performance of the AE method in investigating the behavior of initiation and propagation of interlayer failure and predicting the growth curve of interlayer separation in composite samples [54]. Saeedifar et al. found that damage diagnosis, damage identification and damage localization in multi-layer composites can be detected using the features of the AE method. Also, cracking and breaking of fibers can be detected by having a much higher AE rate than matrix cracks. In some cases, it is possible to differentiate between matrix cracking, delamination and local fiber damage. However, the identification of different fatigue damage mechanisms may be difficult, because the AE signals are often produced by signals caused by friction and wear of the crack surface with each loading. With a careful analysis of the AE spectrum and good sound filter techniques, it is possible to differentiate between certain types of fatigue damage [55].

The research of Šofer et al. showed that the failure of fibers, warping, breaking and cracking of the matrix can be detected using the AE method. Matrix crack failure can be evaluated using the AE test in the frequency band signal between 150 and 200 kHz in carbon fiber composite pipes [56]. Jiang et al. found that the bending behavior and damage evolution based on three-dimensional strain in the reinforced composite can be detected using the AE test [57].

The reliability of the AE method for accurate monitoring of fatigue progress has been proven in numerous studies. **Figure 3** [58] shows the relationship between the amount of AE and strain increase for a boron/epoxy composite during 1 load-controlled fatigue test. As seen in **Figure 3**, the strain gradually increases up to about 1.5 million cycles and then as the composite approaches the final failure, its speed increases faster with more loading. This trend is reflected by the results of AE, which is low at first, but due to the increase in the occurrence of delamination cracking (layering) and fiber breakage, it rapidly progresses toward decrease.

Romhany et al. predicted the fatigue growth curve of interlayer separation in carbon/epoxy samples by the AE method [59]. The results of the study by Mousavi Nasab et al. showed the optimal performance of the AE method in determining the moment of failure in the sample repaired with a composite patch [60].

Impact loading can cause local failure in composite structures, which greatly reduces its strength and stability. These failures are usually internal and cannot be detected by visual inspection. Among the different methods of NDT, AE is of great interest for detecting failure due to its high sensitivity to the processes that stress waves create inside the part [61]. The study of James et al. also showed that the AE method is a suitable method for impact damage ascertainment in composite plates [62]. Jang et al. obtained useful information about low-speed impact events and locations with the AE inspection system and the triangulation method at different sensor distances and impact locations in a composite stiffened plate [63]. In 2014, Zarif Karimi et al. studied the monitoring of reduced strength of perforated composite parts by the AE method. According to the results, the advance rate was identified as the most important factor and the frequency range of matrix failure was 62.5–125 kHz, fiber slippage was 312.5–250 kHz and fiber failure was 375–312.5 kHz [64].

In 2014, Ammar et al. studied the mechanical behavior and AE technique in order to detect defects in sandwich panels. Two sandwich panels with different layers of

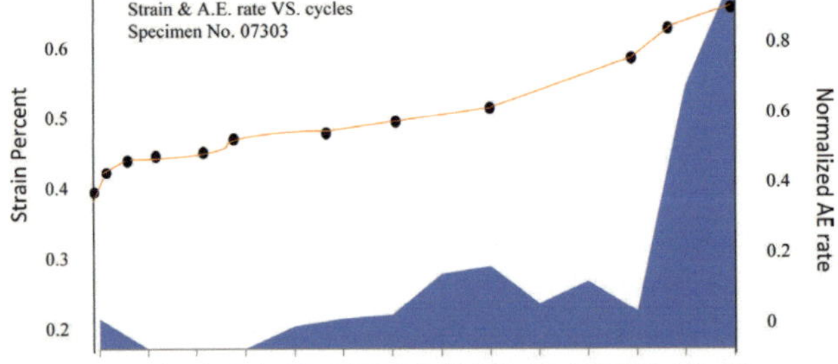


**Figure 3.**
*The effect of the number of tension-tension load cycles on the increase in strain and the amount of AE of a boron/epoxy composite in load control [58].*

porcelain as a shell and foam made of PVC with a density of 60 and 100 kg/$m^3$. In order to create defects in the samples, the foams were subjected to different number of angular cuts, so that the density of defects in them was different. In the meantime, the AE test was also used to detect and characterize the location of defects in two types of sandwich panels during four-point bending loading [65].

In 2014, Ahmadi Najafabadi et al. analyzed and monitored the onset of failure in aluminum/composite joints using the AE method. Also, the moment of onset of failure was detected, the critical force and, accordingly, the fracture toughness were calculated. Modeling was done based on the obtained fracture toughness values [66]. In 2016, Ahmadi Najafabadi et al. investigated the damage monitoring of T3-2024 aluminum sheet repaired with patching of multi-layer metal fibers with the AE method. They divided the samples into 4 groups according to the crack angle of 0 and 45 degrees (repaired and unrepaired state). By using the total energy of AE waves, the onset and critical growth of separation have been identified [67]. In 2007, Wu and Choi used AE with the aim of studying the analysis of the fracture process in slotted multilayer composites with different layers. Their results showed that the final path of the original crack coincides with the orientation of the fibers [68].

Oskouei et al. investigated interlaminar fracture toughness in polyester/glass multilayers subjected to mode I loading by the AE method [69]. Silversides et al. used the AE method to determine the interlaminar fracture toughness in carbon/ epoxy composites subjected to mode I, II and combined mode I and II loading [70]. Allagui et al. investigated the gradual degradation and evolution of damage in the bio-composites during experimental tests using the AE method [71]. For damage area real-time localization of carbon-fiber-reinforced plastics (CFRP) composite laminates, a method based on deep convolutional neural network (CNN) and transfer learning presented using AE signals by Zhao et al. that study showed that the proposed damage area localization method can be applied for SHM of intelligent perception of composite materials [72].

LZ complexity was used to identify the damage modes in a plain weave fabric CFRP specimens tested at an elevated temperature by Barile et al. [73]. This study showed that delamination or debonding can be detected by the AE signals with LZ complexity above 0.6. Open-hole plain woven composites (OHPWCs) under tensile load for damage mode identification, damage initiation detection and damage evolution were analyzed by Liu et al. [74]. Damage participation rate (DPR) based on AE count to characterize the contribution of different damage modes to the overall failure of the plain woven composites presented in this study.

Rubio-González et al. applied the synergistic combination of AE and self-sensing capability provided by the integration of carbon nanotube (CNT) networks to SHM of glass fiber epoxy composites under flexural loading. Damage mechanisms such as matrix cracking, fiber/matrix debonding, delamination and fiber breakage can be identified by this method [75].

Wang et al. applied AE to SHM of the three-point bending process of 3D_C/C_TiC_Cu composites and evaluated the damage change of the materials under different bending stress. The result showed that the AE method is a useful tool for investigating damage evolution of 3D_C/C_TiC_Cu composites [76]. Kucukkalfa et al. applied AE test for damage detection of carbon nanotube (CNT)/ cellulose nanocrystal (CNC)-reinforced foam-cored sandwich composites under flexural load. The test results showed that reinforcements can affect the retardation and/or elimination of core damage, face sheet-core debonding, matrix cracking and fiber breakage in the sandwich composite structures. With the addition of 0.1 wt.% CNT caused the strengthen of core material, as a result, the ratio of AE signals related to fiber breakage and core damage was decreased [77].

Kalteremidou et al. used AE to identify the dominant stress/strain component in carbon/epoxy composite materials even before damage mechanisms became evident. The result showed that AE can applied for identification of the dominant stress/strain component at early loading steps [78]. Yang et al. applied AE to identify the damage mechanism of continuous alumina fiber-reinforced alumina matrix composites ($Al_2O_3f/Al_2O_3$) composites. The results showed that numerous high-energy AE signals were generated with a decreasing of initial load [79]. Wu and Pei applied AE to SHM of carbon fiber-reinforced laminate composites with torsional loads. The test results showed that combining the AE technique, micro-CT and SEM can be a

| **Damage modes** | **Materials** | **The utilized AE parameters** | **Reference** |
|---|---|---|---|
| Matrix damage and delamination | Pultruded fiber-reinforced composite (PFRC) | Rise time, counts, energy, endurance, amplitude, peak frequency | Jiang et al. [57] |
| Delamination | Unidirectional Cytec AS4/5276-1 carbon epoxy prepeg | Energy | Silversides et al. [70] |
| Interlaminar fatigue crack growth | MWCNT/carbon fiber reinforced hybrid composites | Energy | Romhany and Szebényi [59] |
| Weak adhesion | Carbon fiber reinforced polymer (CFRP) | Cumulative energy | Teixeira de Freitas et al. [81] |
| Voids | Glass-fiber-reinforced-polymers (GFRP) | Energy | Kosmann et al. [82] |
| Impact damage | CFRP | Energy | James et al. [83] |
| Delamination | Glass/polyester composites | Energy | Oskouei et al. [69] |
| Delamination | Glass fiber/epoxy composite | Sentry function and frequency | Fotouhi & Najafabad [84] |
| Porosity and fiber waviness | Fiber reinforced composites | Cumulative energy and counts | Qamhia et al. [85] |
| Debonding | Fiber reinforced polymer (FRP)- and steel reinforced grout (SRG) | Cumulative energy | Verstrynge et al. [86] & [87] |
| Crack growth | Aluminum sheet repaired with fiber metal laminate patch | Energy | Ahmadi Najafabadi et al. [67], |
| Debonding | A methacrylate-based universal hybrid composite (Filtek-Z250), a flowable composite (Filtek-Z350 flowable) and a silorane-based composite (Filtek-P90) were investigated. | Cumulative events | Cho et al. [88] |
| Matrix crack, fiber de-bonding, fiber breakage and foam crack | Foam/glass-polyester sandwich panels | Energy | Hamzeloo et al. [53] |

| Damage modes | Materials | The utilized AE parameters | Reference |
|---|---|---|---|
| Matrix cracking | Graphite/epoxy | Energy | Prosser et al. [89] |
| Fiber cracking | Al3Ni fiber reinforced composite | Energy | Harris [90] |
| Crack | Glass-fiber/epoxy laminates | Events | Wu and Choi [68] |
| Crack | Fiber reinforced composite | Event count and energy | Yu et al. [50] |
| Delamination | CFRP | Events, cumulative energy and cumulative events | Barile [73] |
| Interfacial debonding, matrix cracking, fiber breakage and core failure | Sandwich composite with glass/epoxy skin and foam core | Cumulative counts | Pashmforoush et al. [91] |
| Fiber breakage | Carbon/epoxy | Events | Chou et al. [92] |
| Matrix cracking, fiber/matrix, debonding, delamination and fiber breakage | Glass fiber epoxy | Cumulative energy and strength cumulative | Rubio-González et al. [75] |
| Delamination | $Al_2O_{3f}/Al_2O_3$ | Events, cumulative energy | Yang et al. [79] |

**Table 3.**
*SHM of composite structures using the AE method.*

useful method for SHM of composite structures under torsion with different off-axis angle structures [80]. A summary of SHM of composite structures using the AE method is given in **Table 3**.

## 5. Conclusion

SHM of composite structures is an industry requirement. AE method can be used as an effective NDT method to continuously monitor the development of damages in composites. In this chapter of the book, it has been prepared with the aim of application of AE to detect damage in composites. For this purpose, the literature done in the field of SHM of composite structures using the AE method is reviewed. AE is an NDT method to diagnose damage in composites. The capabilities and limitations of this method are described in this chapter. Previous studies have shown that inspection of composites using the AE method can provide valuable insight into failure mechanisms that occur at different stages during fatigue life and ultimately lead to failure.

## Conflict of interest

The author declares no conflict of interest.

## Author details

Sattar Mohammadi Esfarjani
Faculty of Engineering, Department of Mechanical Engineering, University of Zanjan, Zanjan, Iran

*Address all correspondence to: satar.iran@gmail.com

## References

[1] Esfarjani SM, Salehi M. Optimization the inner product vector method and its application to structural health monitoring. Journal of Vibroengineering. 2017;**19**(4):2578-2585

[2] Mohammadi Esfarjani S, Salehi M. Damage identification in aluminium T3-2024 alloy via cross correlation functions. In: The 15th International Conference of Iranian Aerospace, Tehran, Iran. Tehran, Iran: The Faculty of Civil Aviation Industry. 11-13 Mar 2016

[3] Esfarjani SM. Structural damage detection using modal flexibility method in honeycomb composite sandwich beam. Romanian Journal of Acoustics and Vibration. 2020;**17**(1):51-56

[4] Modupalli N, Naik M, Sunil C, Natarajan V. Emerging non-destructive methods for quality and safety monitoring of spices. Trends in Food Science & Technology. 2021;**108**: 133-147

[5] Mohammadi Esfarjani S, Salehi M. Inspection of aboveground pipeline using vibration responses. Journal of Pipeline Systems Engineering and Practice. 2020;**11**(3):04020021

[6] Mohammadi Esferjani S, Salehi M. Evaluation of the damage detection capability of inner product vector for LOP and LOSWF defects in a groove weld. Modares Mechanical Engineering. 2016;**16**(6):7-16

[7] Brunner AJ. Structural health and condition monitoring with acoustic emission and guided ultrasonic waves: What about long-term durability of sensors, sensor coupling and measurement chain? Applied Sciences. 2021;**11**(24):11648

[8] Bøving KG. NDE Handbook: Non-destructive Examination Methods for Condition Monitoring. Butterworth-Heinemann, Elsevier; 12 May 2014. ISBN-10: 1483176533

[9] Dwivedi SK, Vishwakarma M, Soni A. Advances and researches on non destructive testing: A review. Materials Today: Proceedings. 2018;**5**(2):3690-3698

[10] Qu Z, Jiang P, Zhang W. Development and application of infrared thermography non-destructive testing techniques. Sensors. 2020;**20**(14):3851

[11] Towsyfyan H, Biguri A, Boardman R, Blumensath T. Successes and challenges in non-destructive testing of aircraft composite structures. Chinese Journal of Aeronautics. 2020;**33**(3):771-791

[12] Başyiğit AB. A review on non-destructive testing methods for engineering materials. In: Yeliz AŞÇI editor. Engineering Sciences Innovative Approaches. 2021. pp. 191-215. Available from: https://www.bookchapter.org/kitaplar/Engineering%20Sciences%20Innovative%20Approaches_.pdf#page=198

[13] Kah P, Mvola B, Martikainen J, Suoranta R. Real time non-destructive testing methods of welding. Advanced Materials Research. 2 Jul 2014;**933**:109-116. DOI: 10.4028/www.scientific.net/AMR.933.109

[14] Introduction to Acoustic Emission. 2023. Available from: http://www.idinspections.com/acoustic-emission-phenomenon%20/

[15] Gholizadeh S. A review of non-destructive testing methods of composite materials. Procedia Structural Integrity. 2016;**1**:50-57

[16] Grabowski K, Gawronski M, Baran I, Spychalski W, Staszewski WJ, Uhl T, et al. Time–distance domain transformation for acoustic emission source localization in thin metallic plates. Ultrasonics. 2016;**68**:142-149

[17] Rastegaev I, Merson D, Rastegaeva I, Vinogradov A. A time-frequency based approach for acoustic emission assessment of sliding wear. Lubricants. 2020;**8**(5):52

[18] Wang K, Liu X, Wu X, Zhu Z. Monitoring the lack of grease condition of rolling bearing using acoustic emission. International Journal of Modelling, Identification and Control. 2019;**31**(1):94-102

[19] Hase A, Mishina H, Wada M. Acoustic emission in elementary processes of friction and wear: In-situ observation of friction surface and AE signals. Journal of Advanced Mechanical Design, Systems, and Manufacturing. 2009;**3**(4):333-344

[20] Wuriti G, Chattopadhyaya S, Thomas T. Acoustic emission test method for investigation of m250 maraging steel pressure vessels for aerospace applications. Materials Today: Proceedings. 2022;**49**:2176-2182

[21] Toutountzakis T, Mba D. Observations of acoustic emission activity during gear defect diagnosis. NDT & E International. 2003;**36**(7):471-477

[22] El-Ghamry M, Reuben RL, Steel JA. The development of automated pattern recognition and statistical feature isolation techniques for the diagnosis of reciprocating machinery faults using acoustic emission. Mechanical Systems and Signal Processing. 2003;**17**(4):805-823

[23] Barat V, Marchenkov A, Bardakov V, Zhgut D, Karpova M, Balandin T, et al. Assessment of the structural state of dissimilar welded joints by the acoustic emission method. Applied Sciences. 2022;**12**(14):7213

[24] Perveitalov OG, Nosov VV, Borovkov AI, Khanukhov KM, Chetvertukhin NV. Calculation of durability and fatigue life parameters of structural alloys using a multilevel model of acoustic emission pulse flow. Metals. 2022;**13**(1):4

[25] Marasanov V, Stepanchikov D, Sharko A, Sharko O. Technology for determining the residual life of metal structures under conditions of combined loading according to acoustic emission measurements. In: International Conference on Data Stream Mining and Processing, 21-25 August, 2020, Lviv, Ukraine. Cham: Springer International Publishing; 2020. pp. 202-217. DOI: 10.1007/978-3-030-61656-4_13

[26] Rathod VB, Kumbhar GB, Bhalja BR. Partial discharge detection and localization in power transformers based on acoustic emission: Theory, methods, and recent trends. IETE Technical Review. 2022;**39**(3):540-552

[27] Huang M, Jiang L, Liaw PK, Brooks CR, Seeley R, Klarstrom DL. Using acoustic emission in fatigue and fracture materials research. Journal of Metals. 1998;**50**(11):1-14

[28] Khoshouei M, Bagherpour R. Application of acoustic emission (AE) in mining and earth sciences: A review. Rudarsko-geološko-naftni zbornik. Jun 2019;**34**(4):19-32

[29] Michlmayr G, Cohen D, Or D. Sources and characteristics of acoustic emissions from mechanically stressed geologic granular media—A review. Earth-Science Reviews. 2012;**112**(3-4):97-114

[30] Benabdallah HS. Tribological Properties and Acoustic Emissions of Some Thermoplastics Sliding against SAE5 2100. In: Journal of Tribology. Vol. 128. The American Society of Mechanical Engineers - ASME; 2006. pp. 96-102

[31] Zaki A, Chai HK, Aggelis DG, Alver N. Non-destructive evaluation for corrosion monitoring in concrete: A review and capability of acoustic emission technique. Sensors. 2015;**15**(8):19069-19101

[32] AlShorman O, Alkahatni F, Masadeh M, Irfan M, Glowacz A, Althobiani F, et al. Sounds and acoustic emission-based early fault diagnosis of induction motor: A review study. Advances in Mechanical Engineering. 2021;**13**(2):1687814021996915

[33] Anastasopoulos A, Kourousis D, Botten S, Wang G. Acoustic emission monitoring for detecting structural defects in vessels and offshore structures. Ships and Offshore Structures. 2009;**4**(4):363-372

[34] Singh CV. Multiscale Modeling of Damage in Multidirectional Composite Laminates. [Phd thesis] USA: Texas A&M University; 2008. Available from: https://core.ac.uk/download/pdf/147131874.pdf

[35] Belarbi A, Dawood M, Acun B. Sustainability of fiber-reinforced polymers (FRPs) as a Construction material. In: Sustainability of Construction Materials. Elsevier; 2016. pp. 521-538

[36] Smith R. Composite defects and their detection. In: Materials Science and Engineering – Vol. III, no. 1. EOLSS Publications; 2009. pp. 103-143

[37] Djordjevic BB. Nondestructive test technology for the composites. In: The 10th International Conference of the Slovenian Society for Non-destructive Testing. Ljubljana, Slovenia; 1-3 September 2009. pp. 259-265

[38] Maillet E, Baker C, Morscher GN, Pujar VV, Lemanski JR. Feasibility and limitations of damage identification in composite materials using acoustic emission. Composites Part A: Applied Science and Manufacturing. 2015;**75**:77-83

[39] Mouritz A. Non-destructive evaluation of damage accumulation AP Mouritz, RMIT University, Australia. Fatigue in Composites: Science and Technology of the Fatigue Response of Fibre-Reinforced Plastics. 2003:242

[40] Hamstad MA. A review: Acoustic emission, a tool for composite-materials studies. Experimental Mechanics. 1986;**26**:7-13

[41] Kim J-S, Lee K-S, Cho W-J, Choi H-J, Cho G-C. A comparative evaluation of stress–strain and acoustic emission methods for quantitative damage assessments of brittle rock. Rock Mechanics and Rock Engineering. 2015;**48**:495-508

[42] Al-Jumaili SK, Pearson MR, Holford KM, Eaton MJ, Pullin R. Acoustic emission source location in complex structures using full automatic delta T mapping technique. Mechanical Systems and Signal Processing. 2016;**72**:513-524

[43] Falakdin RH, RA, Latifi M. Non-destructive damage evaluation of composites with acoustic emission method. MMEP. 2022;**31**(5):20-31.

[44] Godin N, Huguet S, Gaertner R, Salmon L. Clustering of acoustic emission signals collected during tensile tests on unidirectional glass/polyester composite using supervised

and unsupervised classifiers. Ndt & E International. 2004;**37**(4):253-264

[45] Godin N, Huguet S, Gaertner R. Integration of the Kohonen's self-organising map and k-means algorithm for the segmentation of the AE data collected during tensile tests on cross-ply composites. Ndt & E International. 2005;**38**(4):299-309

[46] Özaslan E, Yetgin A, Acar B, Güler MA. Damage mode identification of open hole composite laminates based on acoustic emission and digital image correlation methods. Composite Structures. 2021;**274**:114299

[47] Muir C, Swaminathan B, Almansour A, Sevener K, Smith C, Presby M, et al. Damage mechanism identification in composites via machine learning and acoustic emission. npj Computational Materials. 2021;**7**(1):95

[48] Fotouhi M, Fotouhi S, Ahmadi M. Acoustic emission based study to monitor the initiation and growth of delamination in composite materials. Modares Mechanical Engineering. 2014;**14**(3):78-84

[49] Fotouhi M, Heidary H, Ahmadi M, Pashmforoush F. Characterization of composite materials damage under quasi-static three-point bending test using wavelet and fuzzy C-means clustering. Journal of Composite Materials. 2012;**46**(15):1795-1808

[50] Yu Y-H, Choi J-H, Kweon J-H, Kim D-H. A study on the failure detection of composite materials using an acoustic emission. Composite Structures. 2006;**75**(1-4):163-169

[51] Chang F, Gordon D, Gardner A. A study of fatigue damage in com-posites by nondestructive testing techniques. Fatigue of Filamentary Composite Materials. 1977;**636**:57

[52] Barile C, Casavola C, Pappalettera G, Kannan VP. Application of different acoustic emission descriptors in damage assessment of fiber reinforced plastics: A comprehensive review. Engineering Fracture Mechanics. 2020;**235**:107083

[53] Hamzeloo SR, Oskouei AR, Zakizadeh AM. Applying acoustic emission to investigate failure mechanisms on bending of polymer-based composite sandwich panels. Journal of Science and Technology of Composites. 2020;**7**(3):1029-1039

[54] Saeedifar M, Ahmadi Najafabadi M, Hoseyni H, Mohammadi R. Investigation of initiation and evolution of delamination in glass/epoxy laminated composites using acoustic emission method. Amirkabir journal of. Mechanical Engineering. 2016;**48**(4):411-422

[55] Saeedifar M, Zarouchas D. Damage characterization of laminated composites using acoustic emission: A review. Composites Part B: Engineering. 2020;**195**:108039

[56] Šofer M, Cienciala J, Fusek M, Pavlíček P, Moravec R. Damage analysis of composite CFRP tubes using acoustic emission monitoring and pattern recognition approach. Materials. 2021;**14**(4):786

[57] Jiang W, Zhang Q, Zhang Y, Guo Z, Tu S-T. Flexural behavior and damage evolution of pultruded fibre-reinforced composite by acoustic emission test and a new progressive damage model. International Journal of Mechanical Sciences. 2020;**188**:105955

[58] Williams R, Reifsnider K. Investigation of acoustic emission during fatigue loading of composite specimens. Journal of Composite Materials. 1974;**8**(4):340-355

[59] Romhany G, Szebényi G. Interlaminar fatigue crack growth behavior of MWCNT/carbon fiber reinforced hybrid composites monitored via newly developed acoustic emission method. Express Polymer Letters. 1 Jul 2012;**6**(7):572-580

[60] Mohammadi R, Ahmadi Najafabadi M, Hosseini TH. Investigation of failure mechanisms of notched aluminum plates repaired with composite patches using acoustic emission method. Modares Mechanical Engineering. 2017;**17**(8):406-412

[61] Pearson MR, Eaton M, Featherston C, Pullin R, Holford K. Improved acoustic emission source location during fatigue and impact events in metallic and composite structures. Structural Health Monitoring. 2017;**16**(4):382-399

[62] James R, Joseph RP, Giurgiutiu V. Impact damage ascertainment in composite plates using in-situ acoustic emission signal signature identification. Journal of Composites Science. 2021;**5**(3):79

[63] Jang B-W, Kim C-G. Acoustic emission source localization in composite stiffened plate using triangulation method with signal magnitudes and arrival times. Advanced Composite Materials. 2021;**30**(2):149-163

[64] Zarif KN. Study of residual tensile strength in drilled composite laminates by acoustic emission. Modares Mechanical Engineering. 2014;**13**(15):169-183

[65] Ammar IB, Karra C, El Mahi A, El Guerjouma R, Haddar M. Mechanical behavior and acoustic emission technique for detecting damage in sandwich structures. Applied Acoustics. 2014;**86**:106-117

[66] Ahmadi Najafabadi M, Sedighi M, Salehi M, Hosseini TH. Analysis and detection of failure in aluminum/E-glass prepreg interface with acoustic emission. Journal of Science and Technology of Composites. 2014;**1**(1):13-22

[67] Ahmadi Najafabadi M, Hosseini Toudeshky H, Sedighi M. Damage monitoring of aluminum sheet repaired with fiber metal laminate patch by acoustic emission. Modares Mechanical Engineering. 2016;**16**(2):1-9

[68] Woo S-C, Choi N-S. Analysis of fracture process in single-edge-notched laminated composites based on the high amplitude acoustic emission events. Composites Science and Technology. 2007;**67**(7-8):1451-1458

[69] Oskouei AR, Ahmadi M. Acoustic emission characteristics of mode I delamination in glass/polyester composites. Journal of Composite Materials. 2010;**44**(7):793-807

[70] Silversides I, Maslouhi A, Laplante G. Interlaminar fracture characterization in composite materials by using acoustic emission. In: 5th International Symposium on NDT in Aerospace. e-Journal of Nondestructive Testing (eJNDT). Singapore: NDT.net; 13-15th Nov. 2013

[71] Allagui S, El Mahi A, Rebiere J-L, Bouguecha A, Haddar M. In-situ health monitoring of thermoplastic bio-composites using acoustic emission. Journal of Thermoplastic Composite Materials. 2023;**08927057231154548**

[72] Zhao J, Xie W, Yu D, Yang Q, Meng S, Lyu Q. Deep transfer learning approach for localization of damage area in composite laminates using acoustic emission signal. Polymers. 2023;**15**(6):1520

[73] Barile C. Innovative mechanical characterization of CFRP by using

acoustic emission technique. Engineering Fracture Mechanics. 2019;**210**:414-421

[74] Liu Y, Zhang L, Li Z, Chen Z, Huang K, Guo L. Investigation on damage evolution of open-hole plain woven composites under tensile load by acoustic emission signal analysis. Composite Structures. 2023;**305**:116481

[75] Rubio-González C, de Urquijo-Ventura MP, Rodríguez-González JA. Damage progression monitoring using self-sensing capability and acoustic emission on glass fiber/epoxy composites and damage classification through principal component analysis. Composites Part B: Engineering. 2023;**254**:110608

[76] Wang X, Xie H, Tong Y, Wang B, Hu H. Three-point bending properties of 3D_C/C_TiC_Cu composites based on acoustic emission technology. Mechanical Systems and Signal Processing. 2023;**184**:109693

[77] Kucukkalfa E, Ghaderiaram A, Yildiz K, Fotouhi M, Asadi A, Cebeci H. Damage detection of CNT/CNC-reinforced foam-cored sandwich composites by acoustic emission tests under flexural load. In: AIAA SCITECH 2023 Forum, Article AIAA 2023-2030. AIAA; 2023. p. 2030. DOI: 10.2514/6.2023-2030

[78] Kalteremidou K-A, Aggelis DG, Van Hemelrijck D, Pyl L. On the use of acoustic emission to identify the dominant stress/strain component in carbon/epoxy composite materials. Mechanics Research Communications. 2021;**111**:103663

[79] Yang X, Kun C, Zhou Z, Wen Z, Zhi-hang P, Chen-cheng S, et al. Damage mechanism of Al2O3f/Al2O3 composites by acoustic emission technology. Ceramics International. 2023;**49**(2):2730-2737

[80] Wu Q, Pei N. Structural health monitoring of carbon fiber-reinforced laminate composites with torsional loads by using acoustic emission. Journal of Materials Science. 2022;**57**(28):13433-13445

[81] Teixeira de Freitas S, Zarouchas D, Poulis J. The use of acoustic emission and composite peel tests to detect weak adhesion in composite structures. The Journal of Adhesion. 2018;**94**(9):743-766

[82] Kosmann N, Karsten J, Schuett M, Schulte K, Fiedler B. Determining the effect of voids in GFRP on the damage behaviour under compression loading using acoustic emission. Composites Part B: Engineering. 2015;**70**:184-188

[83] James R, Joseph R, Giurgiutiu V. Impact damage detection in composite plates using acoustic emission signal signature identification. In: Active and Passive Smart Structures and Integrated Systems XIV. Vol. 11376. SPIE: SPIE - International Society for Optics and Photonics; 27 April - 8 May 2020. pp. 116-133. DOI: 10.1117/12.2559747

[84] Fotouhi M, Najafabadi MA. Acoustic emission-based study to characterize the initiation of delamination in composite materials. Journal of Thermoplastic Composite Materials. 2016;**29**(4):519-537

[85] Qamhia II, Lauer-Hunt EM, El-Hajjar RF. Identification of acoustic emissions from porosity and waviness defects in continuous fiber reinforced composites. Advances in Civil Engineering Materials. 2013;**2**(1):37-50

[86] Verstrynge E, Wevers M, Ghiassi B, Lourenço P. Debonding damage analysis in composite-masonry strengthening systems with polymer-and mortar-based matrix by means of the acoustic emission technique. Smart Materials and Structures. 2015;**25**(1):015009

[87] Verstrynge E, Van Balen K, Wevers M, Ghiassi B, Oliveira DV. Detection and Localization of Debonding Damage in Composite-Masonry Strengthening Systems with the Acoustic Emission Technique. 2016.

[88] Cho N, Ferracane J, Lee I. Acoustic emission analysis of tooth-composite interfacial debonding. Journal of Dental Research. 2013;**92**(1):76-81

[89] Prosser W, Jackson K, Kellas S, Smith B, McKeon J, Friedman A. Advanced waveform-based acoustic emission detection of matrix cracking in composites. Materials Evaluation. 1995;**53**(9):1052-1058

[90] Harris D, Tetelman A, Darwish F. Detection of fiber cracking by acoustic emission. ASTM STP. 1972;**505**:238-249

[91] Pashmforoush F, Khamedi R, Fotouhi M, Hajikhani M, Ahmadi M. Damage classification of sandwich composites using acoustic emission technique and k-means genetic algorithm. Journal of Nondestructive Evaluation. 2014;**33**(4):481-492

[92] Chou H-Y, Mouritz AP, Bannister M, Bunsell AR. Acoustic emission analysis of composite pressure vessels under constant and cyclic pressure. Composites Part A: Applied Science and Manufacturing. 2015;**70**:111-120

Chapter 3

# Composite Materials: Application Specifics for Transport Category Aircraft Wings

*Chunmei Chen, Ryan Randall and Dmytro Tiniakov*


## Abstract

Modern composite materials (CM) have wide applications in airframe structures of transport category aircraft. Their application has an important place in providing the minimum mass of an aircraft structure. The successful implementation of CM in airframe structures is directly related to the creation and improvement of computational methods for assessing their strength and stiffness. The most complex task of CM application is to provide the required level of strength and stiffness of separate structural elements and a whole unit. Currently, many methods have been developed to evaluate the strength of CM structural elements. Much attention is paid to the problems of fracture mechanics of CM. Numerous different criteria for fractures of a composite product have been developed, but the question of their unambiguous selection still remains open. In this situation, in order to make a reasonable choice of this or that criterion, additional computational-experimental research in relation to a specific structure is necessary. In this chapter, some aspects of strength determination of a CM aircraft wing have been analyzed. Some aspects of such evaluation have been considered.


**Keywords:** aircraft, wing, composite materials, strength, stiffness

## 1. Introduction

The development of new technologies and structural materials has been and still is directly related to the progress of civilization (see **Figure 1**) [1, 2]. Quantitative and qualitative progress in the development of our civilization has been made possible by the use of new materials. The aviation industry is a particularly clear demonstration of this thesis. Aerospace is a universally recognized indicator of progress in the development of advanced fields of knowledge. Historically, a great number of the achievements of civilization in science, engineering, and materials science have been closely tied to aviation [1–3].

Up to now, aluminum and its alloys have been the main structural material used in aviation. The combination of strength, lightness, and low cost has made it indispensable in aircraft design [1, 4, 5].

 IntechOpen

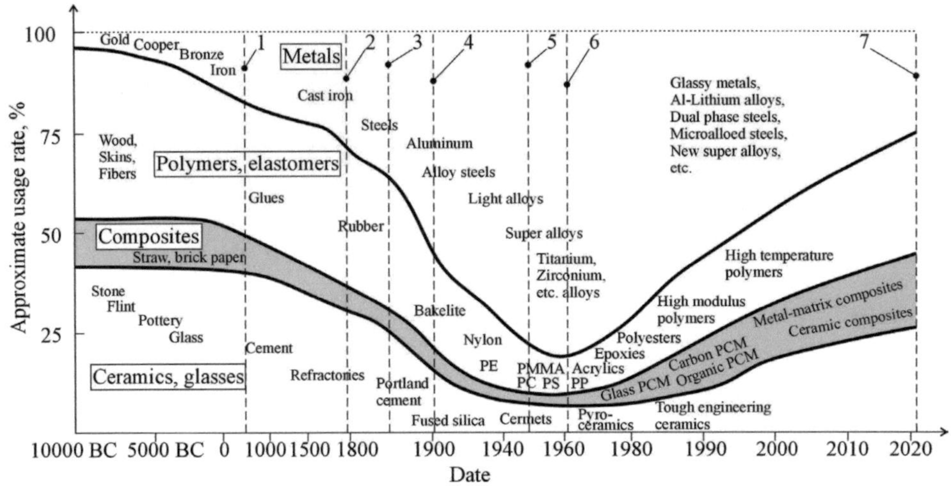


**Figure 1.**
*Developing and evolving structural materials over time [1, 2] (the time scale is nonlinear, graphs drawn on the base of different statistical data). 1: in 559 AD, several prisoners of emperor Wenxuan of Northern Qi, including Yuan Huangtou of Ye, were reportedly forced to fly a kite off a tower as an experiment. 2: the first manned flight, Montgolfier, went aloft in a tethered Montgolfier hot air balloon in 1783. 3: the first manned glider flight was made by a boy in an uncontrolled glider launched by George Cayley in 1853. 4: the first controlled, sustained flight in a powered airplane was made by brothers Wright in 1903. 5: the first documented supersonic flight was by Chuck Yeager in Bell X-1 in 1947. 6: the first piloted orbital flight was made by Yuri Gagarin in the rocket Vostok in 1961. The first manned hypersonic flight was made in 1961 by Robert white in the X-15 research aircraft at speeds in excess of Mach 6. 7: the first powered, controlled takeoff and landing on another planet or celestial body was the NASA rotorcraft ingenuity on Mars in 2021.*

However, progress does not stand still. The ever-increasing demands on aerospace from military and civilian customers have led to a number of processes, one of which has been the search for new structural materials for aerospace industry. In the sixties of the twentieth century, looking for ways to lighten the structure, designers began to use composite materials (CM) everywhere (**Figure 2**), along with new metal alloys [1, 2, 5–7].

Polymer composites are widely used in the aerospace industry. Their composition and manufacturing technology are relatively simple. A woven base in the form of a fabric made of carbon fibers (or other fiber types) is impregnated with polymeric synthetic resins. The raw product is then pressed into the desired shape and applied heat. After the resin has cured, the edges of the product are treated. The product is then typically finished [5, 6].

Composites were first used in aviation in the form of phenol-impregnated modified wood. This material was then replaced by more advanced metal alloys. However, CM returned to aircraft structure when it became necessary to provide radio transparency in the radar antenna zone. Fiberglass composites perfectly met the radio transparency requirements and provided the necessary strength and stiffness as well as aerodynamic perfection for radar fairings [8].

In 1938, the Douglas Aircraft Company used fiberglass for the fairings of the Douglas A-20 Havoc bomber [8]. In 1964, the first all-fiberglass airframe, the H-301 Libelle ("Dragonfly"), received German and U.S.-type certification [8].

After the 1950s, aerospace engineers began to actively develop and introduce new CM reinforced with boron, carbon, and synthetic fibers into aircraft structures [1, 7, 9].

Boeing 727 (1963) was the first medium-range narrow-body airliner developed by the Boeing Corporation to use composite materials in its design. Carbon-epoxy rudder skins were made using CM. A 26% weight reduction of the rudder was achieved [9].

In 1967, the four-seat civilian piston-powered aircraft Eagle, built by Windecker, made its maiden flight in mostly composite [8].

Lockheed L-1011-1 Tristar made its first flight in 1970. Its design used carbon-epoxy aileron skins and later a carbon-epoxy keel. Further models of this aircraft increased the amount of CM. For example, in the Lockheed L-1011-500 Tristar (1978), Kevlar wing-to-fuselage fillets, fixed nose and trailing edge wing panels, trailing edge elevator and rudder panels, wing high-lift fairings, and fuselage fillets with the central engine were used. The weight reduction for CM units was 25% [9].

The design of DC-10 (1970) is generally similar to that of the Lockheed L-1011-1 Tristar. CM were used to make carbon epoxy skins and the rudder structural frame (30% weight reduction compared to metal structure) [9].

Airbus A310 (1982) used a carbon fiber-reinforced plastic vertical stabilizer, control surfaces, high-lift devices, and carbon fiber-based CM brakes (**Figure 3**). The total composite content in the A-310 design was 5% of the aircraft weight [9].

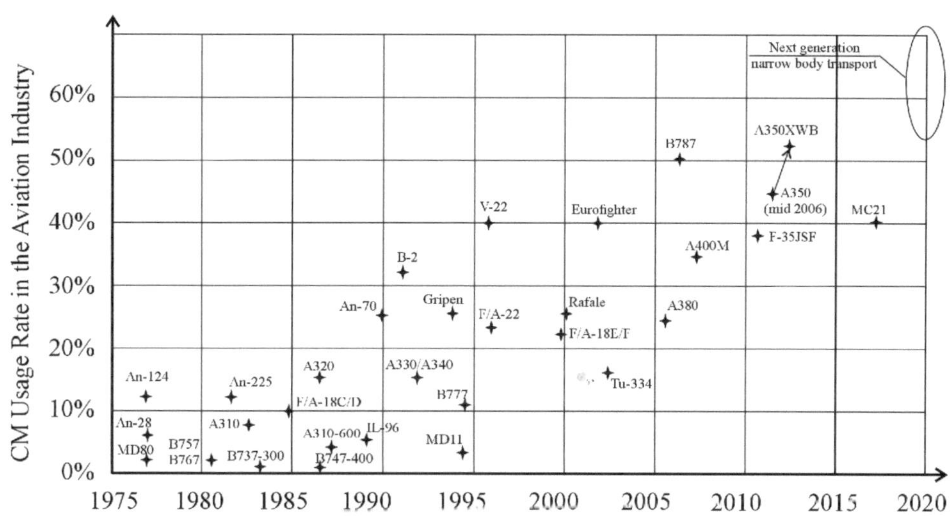


**Figure 2.**
*CM usage rate in the aviation industry [6].*

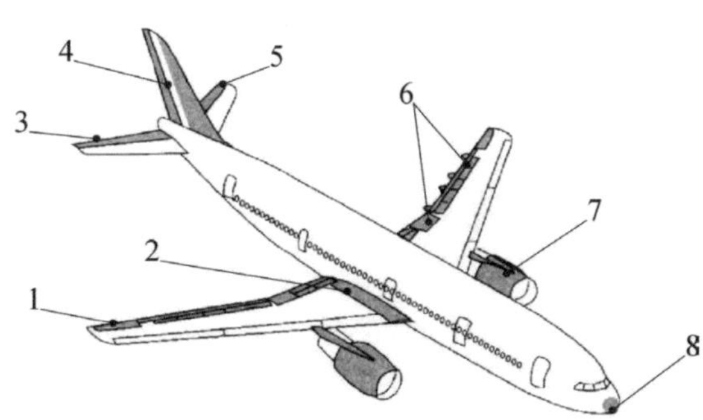


**Figure 3.**
*Composite materials application in Airbus A310 [9, 10]. 1: ailerons; 2: wing-to-fuselage fillets; 3: elevators; 4: rudder; 5: fin; 6: high-lift devices and their fairings; 7: engine case parts; 8: radom.*

ATR-42 (1984) and its extended version, ATR-72 (1989), have wing torsion boxes made entirely of carbon fiber (for the first time in passenger aviation). In general, the proportion of CM in the aircraft was 22.6%. Beech Model 2000 Starship 1 (1986) has wings and fuselage made of CM [8, 9].

Antonov Company (USSR-Ukraine) has a long history of CM application. The share of CM use in Antonov aircraft structures is increasing from 1 to 2% of the airframe weight in An-26 (1969) to 20% in An-70 (1994). Almost all wing high-lift devices were made of CM [9].

The appearance in 2005 of Hawker 4000 (Beechcraft, USA) and European A-380 with an all-composite fuselage brought the percentage of CM in the structure to 25–30% [8, 9].

The development of B787 (2009) and A350 (2013), whose wings and fuselage are mainly made of CM, brought this ratio to 50–55% [1, 4, 9]. Currently, Airbus and Boeing are the undisputed leaders in the use of CM structures in transport category aircraft.

There are also projects to upgrade existing aircraft. The main idea of such upgrading is to replace traditional metal structural materials with composite materials. For example, the Yakovlev Yak-40 upgrade project. CTP-40DT (2016) is a new variant of Yak-40. The wing of this aircraft is completely made of CM [11].

CM are very practical for the UAV industry. Such types of flying vehicles are especially sensitive to their weight. In the past, when structural materials were mainly metals, UAVs had poor flight performance and low applicability. Now, new generation of propulsions and structural materials based on CM use provide high flight performance of UAVs and their wide applications [2, 5, 7, 12–14].

Based on all the above, it can be concluded that the use of composite materials in the design of modern aircraft is reasonable. It is the CM that allow to achieve high weight efficiency for aircraft of the transport category.

## 2. Aircraft wing specifics

The wing is one of the main units of a modern airplane, its main function being to generate lift [1, 4]. However, the wing (as well as the fuselage) is the largest unit of an airplane airframe. This is the cause of high aerodynamic drag. Thus, there is a contradiction: on the one hand, to increase the transport efficiency of the aircraft, the wing area should be increased. On the other hand, an increase in the wing area leads to an increase in its weight, aerodynamic drag, the loads acting on it, etc. [4].

The weight of a passenger aircraft wing is approximately 8–12% of the aircraft structural weight and 30–40% of the airframe weight [4, 15–17]. Up to 30–40% of the structural weight is the weight of the skin. One of the ways to reduce the weight of the wing, which depends on the designer, is the widespread use of composite materials for its elements.

The aerodynamic drag of a wing can be up to 60% of the total drag of an aircraft [4, 15, 18]. This is due to many aspects such as profile drag, friction, and induced drag. In this case, the induced drag is about 30% of the total drag of the aircraft [15, 18] or about 50% of the wing drag. The specificity of the induced drag is that it is directly related to the level of the lift force of the wing [4]. Thus, higher lift force results in higher induced drag.

There are several ways to reduce induced drag for a given lift [4, 11, 15–17]. So far, the most effective way to reduce induced drag is to increase its aspect ratio. For example, the Boeing company used folding wingtips on the new B777X, which allowed the wing aspect ratio to be increased by up to 10 [16, 17] (**Figure 4**). Another example is the new UAC MC-21. This aircraft has an aspect ratio of 11 [17]. In both cases, such a high aspect ratio was achieved by using CM.

One of the consequences of a high aspect ratio is an increase in the bending moment in the wing (for the same lift force) (see **Figure 5**). The increase in internal wing loads leads to an increase in the stresses acting in the wing elements. Therefore, the longer the span of the wing, the more difficult it is to ensure its strength and stiffness. In other words, the increase in mass to provide strength can easily offset any positive effect of using a wing with a large aspect ratio. In addition to bending stiffness, torsional stiffness must also be considered. Torsional stiffness also decreases with increasing wing span.

In this case, the positive property of composite materials, namely, high elasticity, was in full demand. Elasticity is the ability of a material to change its shape under load and return to its original shape when the load is removed. For example, the wing of B787 can bend up to seven meters (**Figure 6c**) [19], and also, the wing of A350XWB can bend up to five meters (**Figure 6d**) [20]. So, where aluminum loses strength and fractures, composites change shape only temporarily, bending under load without breaking.

Another component of the drag of the wing is the joints of the structural elements of the wing that extend into the airflow. The share of such a drag reaches 3% [4, 16, 18, 21].

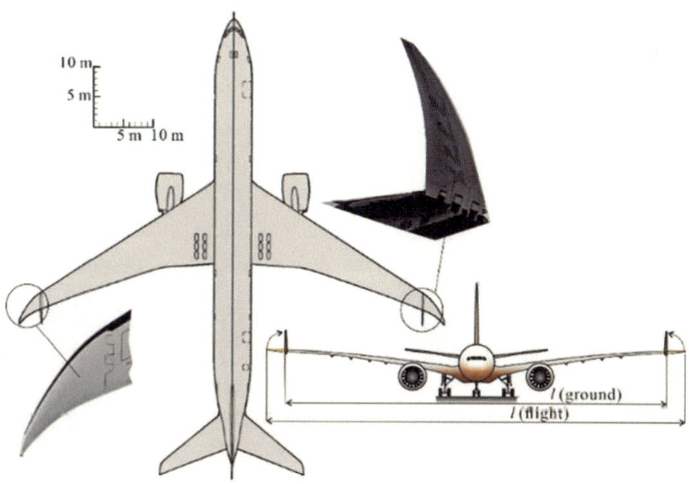


**Figure 4.**
*Deflectable wingtips of Boeing 777X [16].*

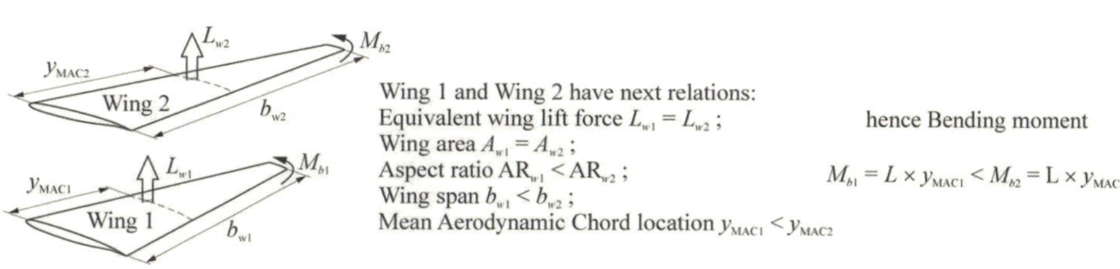


**Figure 5.**
*Equivalent lift force arrangement and bending moment evaluation [1].*

The use of CM reduces the number of joints that extend into the airflow due to the wide use of bonded and solid-state structures.

Thus, the use of composites in the wing structure has a significant positive effect on the main performance indicators of the wing. Namely, it allows to reduce the weight of the wing, reduce its drag by several factors at once, and increase the strength of the wing.

## 3. Composite materials and their performance for aircraft structure

The properties of composite materials depend on the composition of the components, their quantitative ratio, and the strength of the bond between them. By combining the volume content of the components, it is possible to obtain materials with the required values of strength, heat resistance, modulus of elasticity, or compositions with the required special properties, such as high tensile strength, high torsional stiffness, magnetism, and others, depending on the purpose.

**Figure** 7 shows a brief classification of CM by the matrix material type.

Some terms are briefly described in Section 3.5.

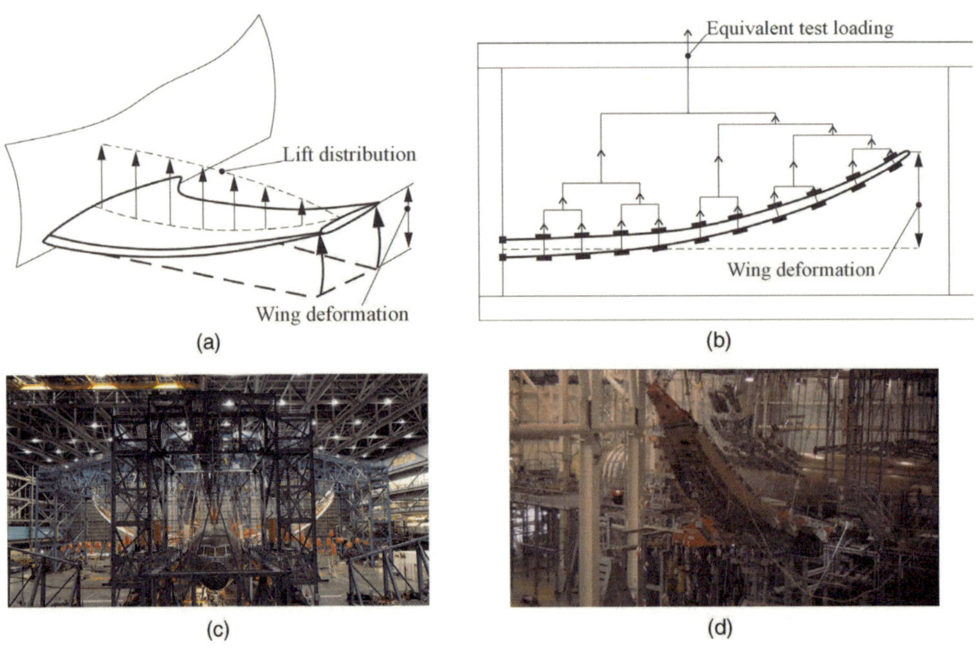


**Figure 6.**
*Wings flex tests. (a) Wing bending scheme in a flight; (b) scheme of the wing bending test; (c) Boeing B787 [19]; (d) Airbus A350XWB [20].*

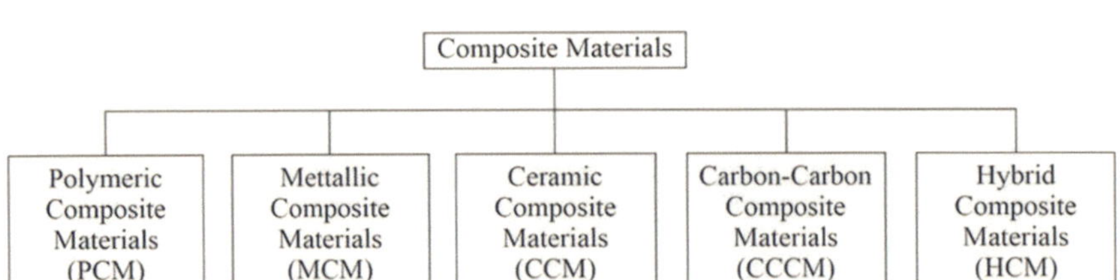


**Figure 7.**
*Brief classification of CM by the matrix material type [6].*

### 3.1 Polymer composite materials

A large group of composites are polymer composite materials (PCM)—composites in which polymer material serves as a matrix. Their use has a significant economic impact.

Parts can be made from PCM using both processes typical of molded polymer products (injection molding, pressing, etc.) and special processes unique to this class of materials (winding, etc.).

#### *3.1.1 Fiberglass*

Fiberglass is a PCM reinforced with glass fibers formed from molten inorganic glass. Thermosetting resins such as polyester, phenolic, epoxy, and others and thermoplastic polymers such as polyamides, polyethylene, polystyrene, and the like are often used as the matrix.

Fiberglass materials have high strength, low thermal conductivity, and high electrical insulation properties and are transparent to radio waves.

Fiberglass is a low-cost polymer composite. Its use is justified in serial and mass production, aerospace industry, shipbuilding, radio electronics, construction, and automotive and railway engineering [2, 5–7, 10].

#### *3.1.2 Carbon fiber-reinforced plastic*

Carbon fiber-reinforced plastics are composite materials consisting of a polymer matrix and reinforcing elements in the form of carbon fibers. Carbon fibers are obtained from synthetic and natural fibers based on copolymers of acrylonitrile, cellulose, and others.

For the production of carbon fiber composites, the same matrices are used for fiberglass—thermosetting and thermoplastic polymers.

The main advantages of carbon fiber composites over glass fiber composites are their low density and higher modulus of elasticity. Carbon fiber-reinforced plastic is a very light and strong material. Carbon fibers, and therefore carbon plastics, have virtually no linear expansion.

Carbon plastics are used in aerospace industry, mechanical engineering, medicine, and sports equipment. Carbon plastics are used to produce high-temperature components for rockets and high-speed aircraft, brake pads and disks for aircraft and reusable spacecraft, and electrothermal equipment [2, 5–7, 10].

#### *3.1.3 Boron fiber-reinforced plastic*

Boron fiber-reinforced plastics are compositions consisting of a polymer matrix and boron fibers. Modified epoxy and polyamide binders are used to make boron plastics. The fibers can be either monofilaments, tapes braided with auxiliary glass filaments, or tapes in which boron filaments are interwoven with other filaments.

Due to the high hardness of the fibers, the material has high mechanical properties, and boron also serves to absorb thermal neutrons.

Boron fiber has high compressive strength, shear strength, hardness, and thermal and electrical conductivity. However, the high brittleness of the material makes it difficult to process and limits the shape of boron plastic products.

Composites based on boron fibers are mainly used in the aerospace industry to produce parts that are subjected to long-term stress [2, 5–7, 10]. The cost of boron fibers is very high due to the peculiarities of their production technology.

#### *3.1.4 Organic fiber-reinforced plastic*

Organic fiber-reinforced plastics are composites of polymeric binders and fillers, which are organic synthetic, less often natural, and artificial fibers in the form of tapes, yarns, fabrics, paper, and others.

In thermosetting organic plastics, the matrix usually consists of epoxy, polyester, and phenolic resins, as well as polyimides. The material contains 40–70% filler. The filler content in organic plastics is based on thermoplastic polymers—polyethylene, PVC, polyurethane, and others—from 2 to 70%.

The degree of orientation of filler macromolecules plays an important role in improving the mechanical properties of organic plastics. Macromolecules of rigid chain polymer (Kevlar) are mainly oriented along the fiber axis and therefore have high tensile strength along the fibers. Body armor is made of Kevlar-reinforced materials. Organic plastics have low density, are lighter than glass and carbon fiber-reinforced composites, have relatively high tensile strength and impact strength but low compressive and flexural strength.

Organic plastics are widely used in automotive, aerospace, radio electronics, shipbuilding, chemical engineering, production of sports equipment, and others [2, 5–7, 10].

### 3.2 Metal matrix composites

Aluminum, magnesium, nickel, copper, and other metals are used to make metal matrix composites (MMC). Fillers are high-strength fibers and refractory particles of varying dispersity that are not dissolved in the base metal.

The properties of dispersion-hardened metal composites are isotropic, that is, the same in all directions. An addition of 5–10% of reinforcing fillers (such as refractory oxides, nitrides, borides, and carbides) leads to increased resistance of the matrix to loading and increased heat resistance of the composite in comparison with the original matrix.

Reinforcing metals with fibers, filamentary crystals, or wires significantly increases both the strength and heat resistance of the metal. For example, aluminum alloys reinforced with boron fibers can be operated at temperatures up to +450–500°C instead of +250–300°C [2, 5–7, 10].

Oxide, boride, carbide, nitride metal fillers, and carbon fibers are used. Ceramic and oxide fibers, due to their brittleness, do not allow plastic deformation of the material, which causes significant difficulties in the manufacture of products, while the use of more plastic metal fillers allows deformation.

Such composites are obtained by impregnating fiber bundles with metal melts, electrodeposition, mixing metal with powder and subsequent sintering, etc. Fiber-reinforced metals are used in aerospace and other industries.

### 3.3 Ceramic matrix composites

Ceramic composite materials (CCM) are materials in which the matrix is ceramic, and the reinforcement is metallic or nonmetallic fillers.

Reinforcing ceramic materials with fibers and dispersed metal and ceramic particles results in high-strength composites.

The range of fibers suitable for reinforcement is limited by the properties of the base material. Metal fibers are often used. The tensile strength does not increase much, but the thermal resistance increases—the material breaks less when heated, but there are cases where the strength of the material decreases. This depends on the ratio of the thermal expansion coefficients of the matrix and the reinforcing fiber. Ceramic composites with carbon fibers are promising for high-temperature applications.

The applications for composite materials are numerous [2, 5, 7, 8, 10, 12–14, 22]. In addition to the aerospace, space, and other specialized industries, they are in demand in the construction of power turbines, the automotive industry, mining, metallurgy, construction, etc. The range of applications of these materials is constantly expanding.

### 3.4 Properties of composite materials

As mentioned above, the main advantages of composite materials are their high specific and fatigue strength, high wear resistance, and stiffness. As the matrix of the composite is responsible for the uniformity of the material, its resistance to external influences, and the distribution and transfer of stresses, the reinforcing material acts as a reinforcing structure and is a stronger component than the matrix. By selecting the properties of the matrix and the reinforcing material, it is possible to achieve the required combination of manufacturing and operational properties. Some performances of different CM are shown in **Table 1** [2, 6, 7, 10, 22].

As shown in **Table 1,** there is a wide range of available properties for CM based on different components. This makes it possible to use CM for different operational conditions (highly stressed elements, thermally stressed elements, radio-transparent elements, etc.) for aircraft structures.

| No. | Matrix material | Reinforcing material | Density, kg/m$^3$ | Tension stress, MPa | Elasticity modulus, GPa |
|---|---|---|---|---|---|
| 1 | Epoxy | Glass fibers | 1900–2200 | 1200–2500 | 50–68 |
| 2 | | Organic fibers | 1300–1400 | 1700–2500 | 75–90 |
| 3 | | Carbon fibers | 1400–1500 | 800–1500 | 120–220 |
| 4 | | Boron fibers | 2000–2100 | 1000–1700 | 220 |
| 5 | Aluminum | Carbon fibers | 2300 | 800–1000 | 200–220 |
| 6 | | Boron fibers | 2600 | 1000–1500 | 220–250 |
| 7 | Magnesium | Carbon fibers | 1800 | 600–800 | 180–220 |
| 8 | | Boron fibers | 2000 | 700–1000 | 200–220 |
| 9 | Nickel | Tungsten wire | 12,500 | 800 | 265 |
| 10 | | Molybdenum wire | 9300 | 700 | 235 |
| 11 | Carbon | Carbon fiber | 1500–1800 | 350–1000 | 120–220 |
| 12 | Ceramic | Silicon carbide fiber | 3200 | 480 | — |

**Table 1.**
*Basic CM properties for different matrix and fiber materials.*

### 3.5 Brief glossary

*Acrylonitrile* is an organic compound with the chemical formula $CH_2CHCN$ and the structure $H_2C{=}CH{-}C{\equiv}N$.

*Aramid* is any of the number of synthetic polymers (substances composed of long-chain, multiunit molecules) in which repeating units containing large phenyl rings are linked together by amide groups.

*Biopolymers* are natural polymers produced by the cells of living organisms.

*Cellulose* is an organic compound with the chemical formula $(C_6H_{10}O_5)_n$ and the most abundant biopolymer on the Earth.

*Electrodeposition* is a process of producing a metal coating on a solid substrate through the reduction of cations of that metal by means of a direct electric current.

*Epoxy resin* is a type of polymer that is made up of a combination of epoxide groups and other molecules.

*Ethylene* is a hydrocarbon which has the chemical formula $C_2H_4$ or $H_2C{=}CH_2$.

*Filamentous crystals* are single crystals in the form of needles and fibers with diameters ranging from a few nanometers to hundreds of microns and a large length-to-diameter ratio (typically 100–1000).

*Kevlar* (para-aramid) is a strong, heat-resistant synthetic fiber, related to other aramids such as Nomex and Technora.

*Monofilament* is a single filament of the synthetic fiber.

*Monomer styrene* is an organic compound with the chemical formula $C_6H_5CH{=}CH_2$.

*Nomex* is a flame-resistant meta-aramid material.

*Para-aramid* is an aromatic polyamide that is characterized by long rigid crystalline polymer chains.

*Phenol* (also known as carbolic acid or phenolic acid) is an aromatic organic compound with the molecular formula $C_6H_5OH$.

*Phenolic resins* are synthetic polymers made by reacting phenol (or substituted phenols) with formaldehyde.

*Polyamide* is a polymer that comprises repeated units linked together by amide bonds.

*Polyester resins* are synthetic resins formed by the reaction of dibasic organic acids and polyhydric alcohols.

*Polyethylene* is a thermoplastic that also happens to be the most common plastic. It is produced through the use of special catalysts that affect polyolefins at moderate temperature and pressure.

*Polyimide* is a polymer containing imide groups belonging to the class of high-performance plastics.

*Polyolefins* are produced by polymerizing, respectively, ethylene and propylene, mainly obtained from oil and natural gas.

*Polystyrene* is a polymer made from the monomer styrene, a liquid hydrocarbon that is commercially manufactured from petroleum.

*Polyurethane* is an ultralight, bearing, elastic, impact-resistant, sound-insulating compound.

*Polyvinyl chloride* (PVC or vinyl) is an economical and versatile thermoplastic polymer.

*Propylene* is a gaseous hydrocarbon with the chemical formula $C_3H_6$, obtained from petroleum.

*Technora* is an aromatic copolyamide that has a highly oriented molecular structure, consisting of both para and meta linkages.

*Thermoplastic* (thermosetting resins) is a type of polymer that liquifies upon heating and cannot be reused after cooling.

## 4. Strength analysis of a composite material wing

### 4.1 Methods for assessing the strength of composite materials

The fracture mechanics of CM units is very complex. There are many approaches to assess their strength based on different fracture criteria [23–25]. They are usually based on experimental results from their tensile, compression, and shear tests. The fracture criterion defines the critical combination of operational stresses (deformations) that will lead to failure.

First, it is necessary to make a conditional partition of the unit into simplified components. For example, for a wing, at least two structural components can be conditionally identified: a panel with stiffeners (skin with stringers) and a spar (**Figure 8**).

For aircraft of the transport category, the intensity of wing loading requires to select the torsion-box load-carrying structure for their wings [4, 26, 27]. The main feature of this type of load-carrying structure is that most of the bending moment is supported by the wing panel. The lift force acts as a shear force into the spar web. So, both of these structural components can be considered as the panel with complex loading configuration. Both these panels have stiffeners: it is a set of stringers for skin and a set of struts for spar web. Caps of a spar can be considered as stiffeners of skin panels. These structural components can also be simplified to a composite package [27].

In the general case, the strength analysis of a composite package is reduced for the determination of the stress-strain state of its layers and the calculation of their safety factors according to one or another criterion [27, 28]. The minimum of these factors determines the safety margin of the CM package as a whole. **Figure 9** shows a typical CM package structure. In the following, the most common fracture criteria used in the practice of composite strength analysis are briefly summarized.

Maximum stress criterion: according to this criterion, monolayer failure occurs when the following conditions are met:

$$A = \max\left(\frac{\sigma_1}{X_T}, \left|\frac{\sigma_1}{X_C}\right|, \frac{\sigma_2}{Y_T}, \left|\frac{\sigma_2}{Y_C}\right|, \left|\frac{\tau_{12}}{S_{12}}\right|\right) \geq 1. \tag{1}$$

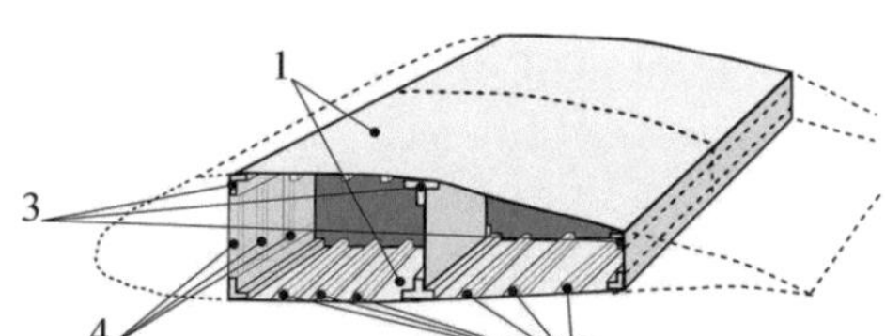


**Figure 8.**
*Wing subparts. 1: upper and lower panels; 2: stiffeners of panels; 3: spars; 4: stiffeners of spars.*

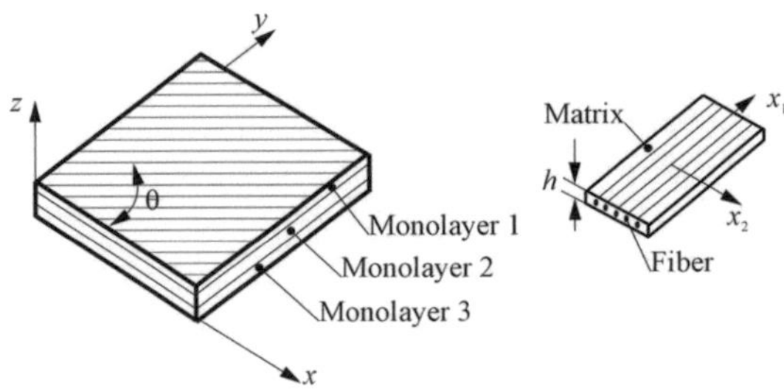


**Figure 9.**
*A typical CM package structure.*

The safety factor is the result of the following equation $\eta = A^{-1}$.

Here, $A$ is the cross-sectional area; $\sigma_1$, $\sigma_2$, and $\tau_{12}$ are the stresses acting in the monolayer; $X_T$, $X_C$, $Y_T$, $Y_C$, and $S_{12}$ are the failure stresses.

This criterion has the following disadvantages. In general, the matrix is in a three-dimensional stress state, but even for relatively thin plates where $\sigma_3 = 0$ and transverse shear strains can be neglected, there is no mutual influence of the components of the plane-stress state on the strength of the matrix, which may lead to overestimation of the matrix strength under a combination of loads.

Tsai-Hill criterion: a quadratic criterion based on the fourth (energy) theory of strength. Monolayer failure occurs when the condition is satisfied:

$$A = \left(\frac{\sigma_1^2}{X^2} - \frac{\sigma_1\sigma_2}{X^2} + \frac{\sigma_2^2}{Y^2} + \frac{\tau_{12}^2}{S_{12}^2}\right) \geq 1. \tag{2}$$

The safety factor here is calculated by the equation $\eta = \sqrt{A}^{-1}$.

The main disadvantage of this criterion is that it is impossible to determine the cause of monolayer failure: matrix or fiber failure has occurred. In addition, the criterion does not distinguish between stress combinations $\sigma_1$ and $\sigma_2$, as biaxial tension or biaxial compression are equivalent in this case.

Hofmann criterion: this criterion is based on the sum of the linear and quadratic stress invariants. Monolayer failure occurs when the following condition is satisfied:

$$A = \left(\frac{\sigma_1^2}{X_T X_C} - \frac{\sigma_1\sigma_2}{X_T X_C} + \frac{\sigma_2^2}{Y_T Y_C} + \frac{\tau_{12}^2}{S_{12}^2} + \sigma_1\left[\frac{1}{X_T} - \frac{1}{X_C}\right] + \sigma_2\left[\frac{1}{Y_T} - \frac{1}{Y_C}\right]\right) \geq 1. \tag{3}$$

The safety factor is determined by solving the quadratic equation $\eta^2 + b\eta - 1 = 0$.

The criterion distinguishes between combinations of stresses $\sigma_1$ and $\sigma_2$ due to additional linear terms compared to Hill's criterion. As with Hill's criterion, it is rather problematic to determine the cause of failure.

Tsai-Wu criterion: it is a variation of the Hofmann criterion. Monolayer destruction occurs when the following condition is satisfied:

$$A = \left(\frac{\sigma_1^2}{X_T X_C} + 2F_{12}\sigma_1\sigma_2 + \frac{\sigma_2^2}{Y_T Y_C} + \frac{\tau_{12}^2}{S_{12}^2} + \sigma_1\left[\frac{1}{X_T} - \frac{1}{X_C}\right] + \sigma_2\left[\frac{1}{Y_T} - \frac{1}{Y_C}\right]\right) \geq 1. \tag{4}$$

This criterion differs from the Hofmann criterion by the factor used for $\sigma_1$ and $\sigma_2$. In general, the factor $F_{12}$ can be in the range $-1.0 \leq F_{12} \leq 1.0$.

A characteristic feature of the criteria discussed above is their uncertainty with respect to what happened in the monolayer—whether the matrix or the fiber was destroyed. At the same time, this is very important when analyzing the strength of a composite package. Therefore, criteria that analyze the strength margins of both the matrix and the fiber separately are becoming more common.

Hashin-Rotem criterion: this criterion assesses fiber and matrix strength separately.

The following relationships are used to determine the fiber strength:

$$\begin{aligned} &\left(\frac{\sigma_1}{X_T}\right)^2 + \left(\frac{\tau_{12}}{S_{12}}\right)^2 = 1 \quad \text{for } \sigma_1 > 0, \\ &\left(\frac{\sigma_1}{X_C}\right)^2 = 1 \quad \text{for } \sigma_1 < 0. \end{aligned} \tag{5}$$

The strength of the matrix is determined by the following relations:

$$\begin{aligned} &\left(\frac{\sigma_2}{Y_T}\right)^2 + \left(\frac{\tau_{12}}{S_{12}}\right)^2 = 1 \quad \text{for } \sigma_2 > 0, \\ &\left(\frac{\sigma_2}{2S_{23}}\right)^2 + \left[\left(\frac{Y_C}{2S_{23}}\right)^2 - 1\right]\frac{\sigma_2}{Y_C} + \left(\frac{\tau_{12}}{S_{12}}\right)^2 = 1 \quad \text{for } \sigma_2 < 0. \end{aligned} \tag{6}$$

The conditions for interlayer delamination are specified by the relation:

$$\left(\frac{\sigma_3}{Z_T}\right)^2 + \left(\frac{\tau_{23}}{S_{23}}\right)^2 + \left(\frac{\tau_{31}}{S_{31}}\right)^2 = 1. \tag{7}$$

The criteria considered above belong to the group of criteria based on the ultimate stresses of the monolayer. In addition, there are a number of criteria based on the ultimate strain of the monolayer.

The maximum strain criterion is one of the most common criteria. According to this criterion, monolayer failure occurs when the following conditions are met:

$$A = \max\left(\frac{\varepsilon_1}{\varepsilon_{1T}}, \left|\frac{\varepsilon_1}{\varepsilon_{1C}}\right|, \frac{\varepsilon_2}{\varepsilon_{2T}}, \left|\frac{\varepsilon_2}{\varepsilon_{2C}}\right|, \left|\frac{\varepsilon_{12}}{\gamma_{12}}\right|\right) \geq 1. \tag{8}$$

The safety factor is calculated using the following equation $\eta = A^{-1}$.

Here, $\varepsilon_1$, $\varepsilon_2$, and $\varepsilon_{12}$ are the strains of the monolayer; $\varepsilon_{1T}$, $\varepsilon_{1C}$, $\varepsilon_{2T}$, $\varepsilon_{2C}$, and $\gamma_{12}$ are the strains and their failure in tension, compression, and shear. Formally, this criterion is the same as the maximum stress criterion. The difference between them is that the elastic characteristics of the matrix can have a nonlinear behavior, and the strain criteria make it possible to take this factor into account to a certain extent.

There are other criteria with different approaches to taking into account stresses and strains. Here, the more commonly used one has been explained.

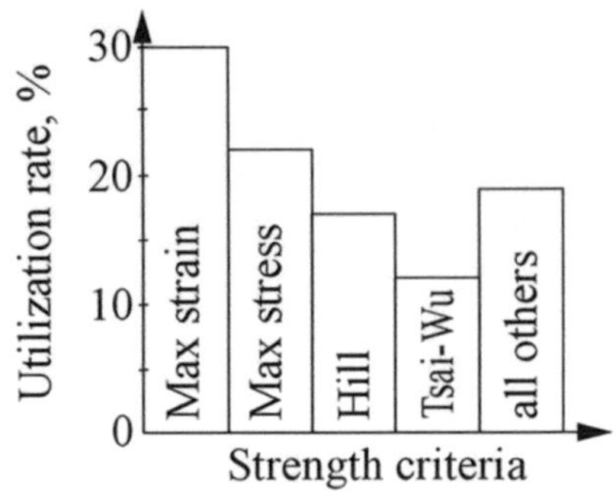


**Figure 10.**
*Application of various failure criteria to analyze the strength of composite units [28].*

**Figure 10** illustrates the extent to which certain strength criteria are used in practice.

## 4.2 Strength and stability analysis of CM panels with stiffeners

Depending on the stiffness ratio of skin and stiffeners, the reinforced CM plates can be considered as composite plate sets connected along nodal lines or as structurally orthotropic sets with given stiffness characteristics. Depending on the model adopted, different numerical approaches are also used for the strength analysis.

The study with discrete models allows detection, in addition to the general instability, of some local forms of loss of stability. However, to fully study the behavior of the panel after a local loss of stability, a discrete model should be used in combination with geometric and physical nonlinearities.

Therefore, when solving the problem in a nonlinear form, the nonlinear stress-strain behavior should be analyzed at each step according to the strength criteria in order to determine the possible failure of the plate before a general form of loss of stability occurs. This type of failure is called crippling strength in the reference literature.

In addition to the abovementioned complex way of studying the behavior of a reinforced panel based on a discrete model in a geometric nonlinear form, it is possible to use a model of a structurally orthotropic plate as a substitute for a real reinforced panel. In this case, the problem is reduced to the calculation of the "smeared-out" stiffness parameters of such an orthotropic plate.

### *4.2.1 Numerical-analytical method for calculation of the overall stability of CM panels, taking into account the local loss of skin stability*

The solution of the general stability problem of a reinforced CM panel is reduced to the solution of the stability problem of an anisotropic (orthotropic) plate based on the equation:

$$D_x \frac{\partial^4 w}{\partial x^4} + 2\left(\mu_{xy} D_x + 2D_{xy}\right) \frac{\partial^4 w}{\partial x^2 \partial y^2} + D_y \frac{\partial^4 w}{\partial y^4} = -N_x^o \frac{\partial^2 w}{\partial x^2} - 2N_{xy}^o \frac{\partial^2 w}{\partial x \partial y} - N_y^o \frac{\partial^2 w}{\partial y^2}. \quad (9)$$

Here, $w$ is the lateral deflection; $D_x$, $D_y$, and $D_{xy}$ are the stiffnesses along their respective axes; $N_x$, $N_y$, and $N_{xy}$ are the forces in their mid-plane.

The critical compressive forces can be approximated by the following equations:

1. A rectangular plate hinged along the contour, loaded at the edges $x = 0$ and $x = a$ by the compressive forces $N_x^o$:

$$N_{xcr} = \frac{2\pi^2}{b^2}\left[\sqrt{D_x D_y} + \mu_{xy} D_x + 2D_{xy}\right]. \tag{10}$$

2. Rectangular elongated plate fixed on all edges:

$$N_{xcr} = \frac{2\pi^2}{b^2}\left[4.6\sqrt{D_x D_y} + 2.67\mu_{xy} D_x + 5.33D_{xy}\right]. \tag{11}$$

3. Rectangular elongated plate, one unloaded edge is free and the other three are hinged ($a/b > 4$):

$$N_{xcr} = \frac{12D_{xy}}{b^2} + \frac{\pi^2 D_x}{a^2}. \tag{12}$$

4. Rectangular plate ($1 < a/b < \infty$) whose loaded edges are hinged and whose unloaded edges are fixed:

$$N_{xcr} = \frac{\pi^2}{b^2}\left\{D_x m^2\left(\frac{b}{a}\right)^2 + 2.67\mu_{xy} D_x + 5.33\left[D_y\left(\frac{a}{b}\right)^2 + \frac{1}{m^2} + D_{xy}\right]\right\}. \tag{13}$$

In the practice of studying the stability of reinforced metal panels and shells, the concept of a "framed" shell or panel has developed. This means that the main material is concentrated in the reinforcements and the skin is comparatively thin. The same assumption applies to CM panels. The comparatively early local loss of stability of the skin and the reduction of its performance in the subcritical area are accounted for by the reduction factor method.

The assumption that the shell or panel is "framed" leads to the assumption that there is no local loss of stability of the reinforcement elements. As a result, the considered approach does not take into account the possibility of failure of the reinforcement after local loss of stability.

For a reinforced plate considered as an orthotropic plate, the following relations are valid:

$$\begin{gathered}\{N\} = [B]\{\varepsilon\},\ \ \{M\} = [D]\{\chi\} \\ \left([B] = \left[B_{ij}\right] = [B]^{\mathrm{T}},\ \ [D] = \left[D_{ij}\right] = [D]^{\mathrm{T}},\ \ i,j = 1,2,3\right)\end{gathered}, \tag{14}$$

for $B_{13} = B_{23} = D_{13} = D_{23} = 0$, and $\{N\} = \left[N_x, N_y, N_{xy}\right]^{\mathrm{T}}$ and $\{M\} = \left[M_x, M_y, M_{xy}\right]^{\mathrm{T}}$ are forces and moments in the panel; $\{\varepsilon\} = \left[\varepsilon_x, \varepsilon_y, \gamma_{xy}\right]^{\mathrm{T}}$ and $\{\chi\} = \left[\chi_x, \chi_y, \chi_{xy}\right]^{\mathrm{T}}$ are strains and curvatures at the reference plane.

The following ratios apply to orthotropic skins

$$B_{11_0} = \frac{E_x h}{1 - \mu_{xy}\mu_{yx}}, \quad B_{22_0} = \frac{E_y h}{1 - \mu_{xy}\mu_{yx}}, \quad B_{12_0} = \frac{\mu_{yx} E_x h}{1 - \mu_{xy}\mu_{yx}}, \quad B_{33_0} = G_{xy} h,$$
$$D_{11_0} = \frac{E_x h^3}{12\left(1 - \mu_{xy}\mu_{yx}\right)}, \quad D_{22_0} = \frac{E_y h^3}{12\left(1 - \mu_{xy}\mu_{yx}\right)}, D_{12_0} = \mu_{yx} D_{11_0}, \quad D_{33_0} = \frac{G_{xy} h^3}{12}. \tag{15}$$

If the skin is regularly supported by longitudinal and lateral stiffeners installed with pitches $b_c$ and $b_f$, respectively, the "smeared-out" stiffness parameters of the supported panel are calculated using the following equations:

$$\begin{gathered} B_{11} = E_s\left(r_{ts} h + \frac{F_s}{b_s}\right), \quad B_{22} = E_f\left(r_{tf} h + \frac{F_f}{b_f}\right), \\ B_{33} = G_{xy} h + G_{sf} h_{sf}, \quad B_{12} = B_{12_0}, \quad B_3 = B_{33_0} + B_{12_0}, \\ D_{11} = E_s\left[r_{ts}\left(\frac{h^3}{12} + h h_1^2\right) + \frac{I_s}{b_s} + \frac{F_s}{b_s}(z_s - h_1)^2\right], \\ D_{22} = E_f\left[r_{tf}\left(\frac{h^3}{12} + h h_2^2\right) + \frac{I_f}{b_f} + \frac{F_f}{b_f}(z_f - h_2)^2\right], \\ D_{33} = G_{xy}\frac{h^3}{6}, \quad D_3 = 2D_{33_0} + D_{12_0} + \frac{G_s I_s^P}{2b_s} + \frac{G_f I_f^P}{2b_f}. \end{gathered} \tag{16}$$

Here, $r_{ts} = \frac{E_x}{E_s\left(1-\mu_{xy}\mu_{yx}\right)}$ and $r_{tf} = \frac{E_y}{E_f\left(1-\mu_{xy}\mu_{yx}\right)}$; $E_s$, $E_f$, $G_s$, and $G_f$ are the elastic and shear moduli of the longitudinal and lateral reinforcement; $F_s$, $F_f$, $I_s$, $I_f$, $I_s^P$, and $I_f^P$ are the areas, eigencentral, and polar moments of inertia for the longitudinal and lateral reinforcement sections, respectively; $z_s$ and $z_f$ are the distances from the centers of gravity of the longitudinal and lateral reinforcement sections to the midplane of the skin; $G_{sf}h_{sf}$ is the shear stiffness of the reinforcement set without skin, which is related to the way the longitudinal and lateral reinforcements are attached to each other.

The given equations for the secant and tangent stiffness parameters are further used in the calculation of the precritical stress-strain behavior and the overall stability of the reinforced panel. In the particular case of no skin buckling, these equations are transformed into the usual equations for the stiffness parameters of the reinforced panel.

#### *4.2.2 Methodology for calculating the structural capacity of panels using FEM*

The analysis of the structural capacity of reinforced panels based on FEM requires special approaches compared to, for example, the analysis of smooth plates. The solution of such problems requires a certain experience in their simulation and in analyzing the obtained data. Some peculiarities and tips for solving the problems of assessing the structural capacity of reinforced panels are presented below.

The analysis of panels operating in tension does not cause significant difficulties in assessing their strength. The main difficulties lie in the analysis of the panel operating in compression or under the action of a combination of loads and are related to the issues of its local and overall stability. A peculiarity of reinforced thin-walled structures, which include stiffened panels, is that they can have local forms of stability loss related to the loss of stability of the skin and elements of the reinforcement set. This does not exhaust the structural capacity of the panel, but only reduces its stiffness characteristics. Panel failure occurs either when the load reaches the value of its total loss of stability or when the strength of its separate elements is exhausted.

For metal panels, a natural limitation on the level of development of subcritical strains is the requirements of aviation regulations for the absence of residual strains in the structure under operational loads.

For panels made of CM, these requirements are transformed into requirements of no violation of composite continuity under these loads, that is, no matrix failure or delamination in it. For panels with a comparatively weak set of stiffeners, in addition to the loss of stability of the skin, there may be a local loss of stability of the stiffener elements: their webs or caps, while the stiffener itself remains in the plane of the panel. This phenomenon is known as panel creep and must be taken into account when assessing the structural capacity of the panel. Panel creep can be related not only to the local loss of stability of the stiffener, but also to its deformation interaction with the skin that has lost stability (**Figure 11**).

In composites, creep can quickly lead to stiffener failure. This depends mainly on the specific thickness of the stiffeners, that is, the ratio of wall or cap width to thickness. The higher the ratio, the faster the creep failure. As an example, **Figure 12** shows a graph of the change in the safety factor of a plate made of CM with a ratio of $b/\delta$ = 27 during its subcritical deformation. It can be seen that, after the loss of plate stability, its failure occurs in the 5% range of load increase, and the safety factor accordingly decreases more than twice.

The finite element model of a panel is created based on its geometric data. Typically, this is a bend-diaphragm element. In terms of length, it is desirable to consider three-span panels. This reduces the influence of boundary conditions at the panel edges on the solution results and forms natural boundary conditions for the offset part of the model, that is, the middle span. The lateral dimensions of the model are usually chosen based on the wide panel concept, when there are no natural boundary conditions at the longitudinal edges due to the operation of the panel as part of the structure. The wide panel concept assumes that the boundary conditions at its longitudinal

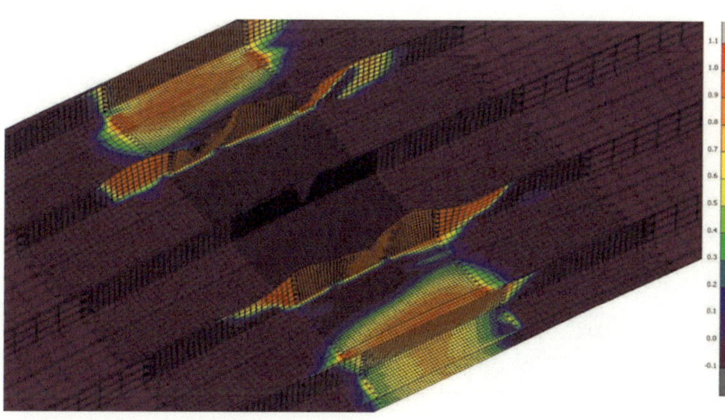


**Figure 11.**
*Deformation interaction of the skin with the set of stiffeners [29].*

edges have minimal influence on the solution. Typically, five to six stiffeners in the model are sufficient to satisfy these conditions using finite elements.

Plate edge torsion should be locked when a fixed state is formed. Lateral movement of the edges should not be restrained. This causes biaxial compression of the skin in the edge zone when the plate is loaded, resulting in an earlier loss of stability. Forming hinge support conditions does not require any additional conditions on the displacements of the edges, except for limiting their displacements out of the panel plane.

The panel loading is provided by setting distributed loads on its edges. The values of these loads are set in proportion to the longitudinal stiffnesses of the panel elements. This results in its central loading.

The size of the computational mesh of the model is mainly determined by the parameters of the panel and the objectives of the analysis. If it is known that there will be no local forms of stability loss, the nodal mesh can be less detailed.

A typical example of such a model of a panel with stiffeners is shown in **Figure 13**.

The application of the nonlinear form of FEM to the analysis of the stability of thin-walled units opens much wider possibilities for a complete analysis of the structural capacity of units. This is mainly due to the possibility of considering the subcritical behavior of a unit and the possibility of monitoring its strength by some or other failure criteria at each stage of its loading. However, there are serious difficulties in the practical use of this approach. First of all, it takes a lot of time to analyze such problems.

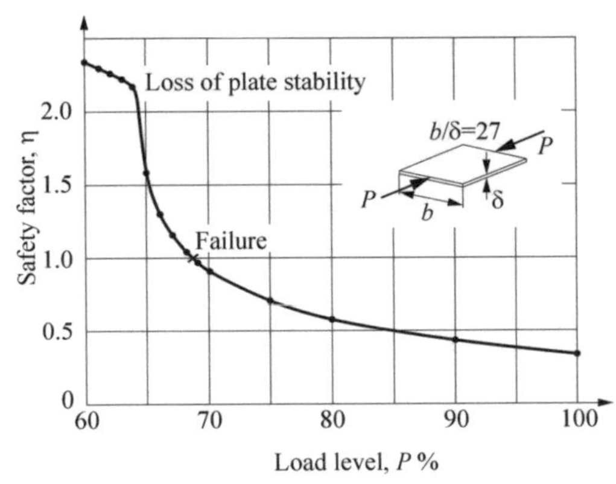


**Figure 12.**
*CM plate safety factor changes after the loss of stability [28].*

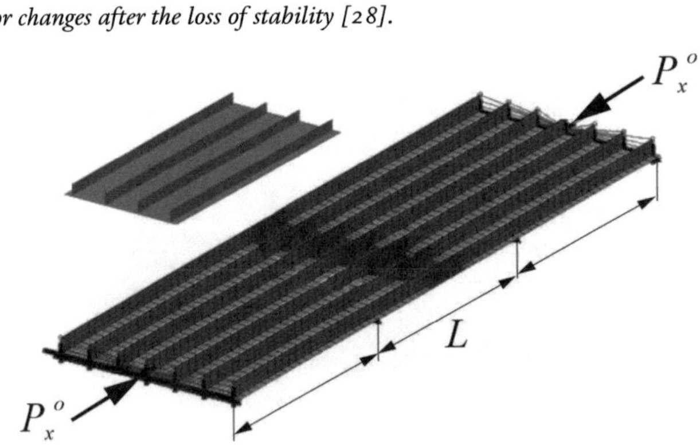


**Figure 13.**
*Typical reinforced panel model for analysis [29].*

The results of the analysis of the reinforced panel show the deformation of the stiffeners from the plane of the panel and the growth of the stresses in the stiffener as a function of the percentage of the load are observed. The intensive growth of the stiffener deformation indicates that the point of total loss of stability of the panel is near, and the analysis of its stress-strain behavior allows to specify the failure load. **Figure 14** shows typical dependencies of stiffener deformation growth and stresses in its rib as a function of load.

The prediction of the critical load can be based on the method of generalized tangential stiffness of the panel using. Conventionally, this stiffness can be defined as the ratio of the percentage of the load increment to the increment of the panel edges convergence value, $T = \Delta P/\Delta U$. As it is the tangential stiffness that determines the stability of the panel, if it decreases sharply, it can be concluded that the critical point of the total loss of panel stability is near. As an example, **Figure 15** shows a typical graph of the convergence of the edges of one of the panels as a function of the percentage of the applied load.

Thus, in the area of 80% load, $T = (87.5 - 75)/(3.97 - 3.148) = 15.2$, and at the finish of the process, $T = (88.67 - 88.65)/(4.133 - 4.088) = 0.44$, that is, the tangential compressive stiffness of the panel decreased almost 35 times.

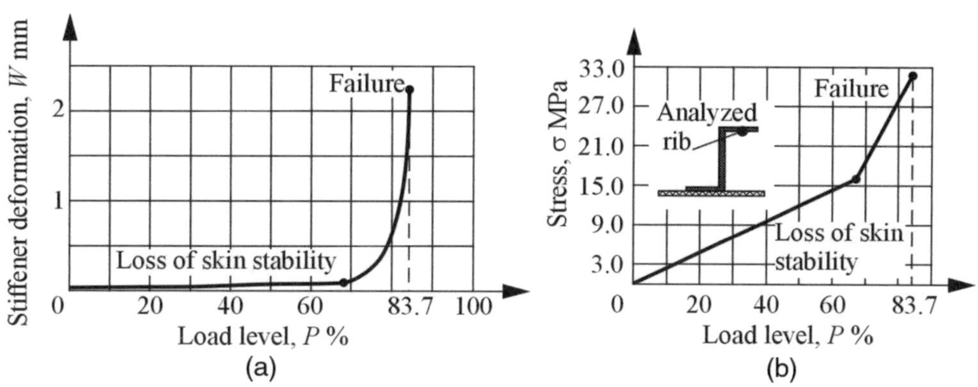


**Figure 14.**
*Graphs of the stringer deformation growth (a) and the stringer rib stress (b) as a function of the load [28, 30].*

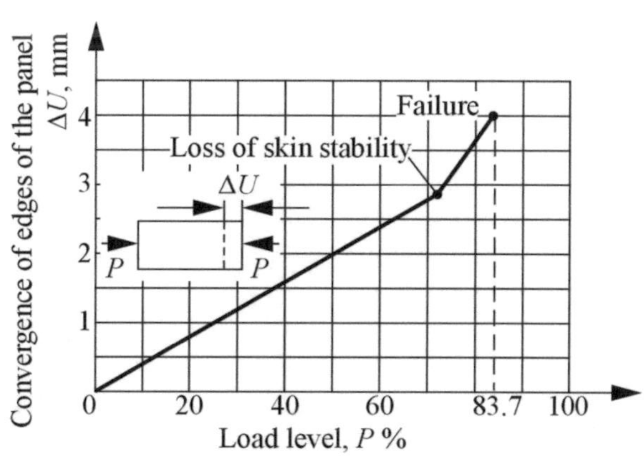


**Figure 15.**
*Graph of the dependence of the value of the panel edges convergence on the percentage of the applied load [28, 30].*

#### *4.2.3 Assessment of panel structural strength*

In the airframe structure, panels typically operate under combined loading conditions, the major components of which are axial and shear loads. The values of these loads depend on both the location of the panel in the structure and the design loading case. Therefore, it is convenient to assess the structural strength of panels using diagrams of their strength and general stability, plotted in the coordinates of the acting linear compressive and shear loads. The combination of these diagrams provides a visual illustration of the structural strength of the panel and the causes of its failure (**Figure 15**).

Presenting such diagrams for a number of typical panels of the considered structure and plotting on them the points of specific combinations of loads acting on the panel in various design cases, it is possible to evaluate its structural strength and available safety factor. At the preliminary stage of design, such diagrams can be drawn using numerical and analytical calculation methods. However, at the stage of detailed design and verification analysis, it is desirable to use nonlinear FEM for this purpose. This is especially true for CM panels. This is due to the fact that analytical calculation methods often do not take into account the effects of panel crippling. As an example, **Figure 16** shows the diagram of the structural strength of a wing panel made of CM as plotted by the results of nonlinear FEM analysis.

The diagram clearly shows that the structural strength of the panel is determined by its overall stability when the ratio of shear and compression loads $\xi = N_{xy}/N_x < 0.32$ and by the strength of the composite when $\xi > 0.32$. Having a point on the diagram that defines the combination of loads acting on the panel, it is easy to estimate the available reserves of its strength by drawing a ray through this point from the origin to its intersection with the boundary of the structural strength of the panel.

## 5. Conclusions

Modern level of material science allows to replace traditional metallic structural materials by new composite materials. This trend provides weight reduction of

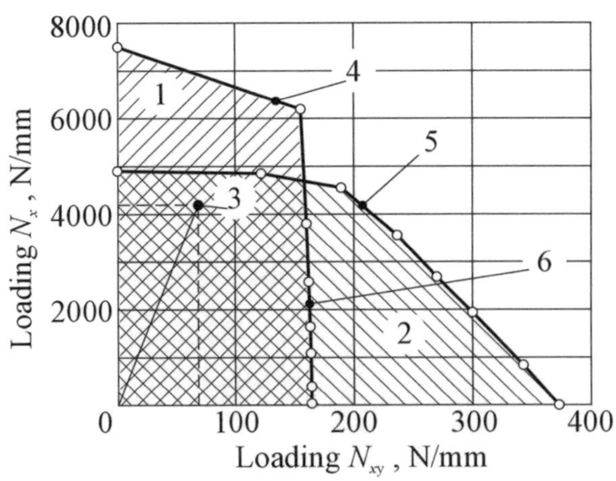


**Figure 16.**
*Diagram of the structural strength of a CM wing panel [28]. 1: zone of the overall instability; 2: zone no strength; 3: design load combinations; 4: limit of strength; 5: limit of stability; 6: limit of structural strength of the panel.*

airframes. However, CM units require more complex approaches to the design, analysis, and verification processes.

There are several criteria to assess the strength of CM basic elements. The aircraft wing structure consists of complex units. For them, it is necessary to apply certain levels of approximation because the full process of strength assessment can take a long time.

The aircraft wing consists of two main structural components: skin and spar. Both can be considered as a reinforced panel. Such an approach allows to unify different analysis models for a relatively well-known model.

On the basis of these conditions, it is possible to assess the strength of the wing by criteria for a reinforced panel. Some peculiarities and recommendations for the selection of parameters of the model of such a panel have been given. Also, several cases have been explained and analyzed as examples.

## Author details

Chunmei Chen[1]*, Ryan Randall[1] and Dmytro Tiniakov[2]

1 Qingdao University, Qingdao, China

2 Nanjing University of Aeronautics and Astronautics, Nanjing, China

*Address all correspondence to: chunmei.chen@qdu.edu.cn

## References

[1] Tiniakov D, Li L. Airworthiness Engineering Basis. Beijing: Beijing Press; 2020. 217 p. ISBN 9787512432093

[2] Ashby M. Material Selection in Mechanical Design. Oxford: Butterworth-Heinemann; 2000. 534 p. ISBN 0750643579

[3] Anderson JD. The Airplane, a History of its Technology. Reston, VA: American Institute of Aeronautics and Astronautics; 2002. 369 p. ISBN 1563475251

[4] Torenbeek E. Advanced Aircraft Design: Conceptual Design, Analysis, and Optimization of Subsonic Civil Airplanes. Chichester, West Sussex: John Wiley and Sons.; 2013. DOI: 10.1002/9781118568101

[5] Breuer UP. Commercial Aircraft Composite Technology. Switzerland: Springer; 2016. 257 p. DOI: 10.1007/978-3-319-31918-6

[6] Li L, Tiniakov D. Airworthiness Design of Composite Structures. Beijing: Beijing press; 2021. 220 p. ISBN 9787512435773

[7] Baker A, Dutton S, Kelly D. Composite Materials for Aircraft Structures. Reston: AIAA Education Series.; 2004. 569 p. ISBN 1563475405

[8] Feygenbaum Y, Butushin S, Bozhevalov D, Sokolov Y. Composite materials and history their introduction in aircraft structures. Scientific Bulletin of GA. 2015;7(318):24-37

[9] Kiva D. Stages of formation and beginning of expand use of polymer composite materials in constructions of passenger and transport airplanes. Aerospace Technic and Technology. 2014;**6**:5-16

[10] Gay D, Hoa SV, Tsai SW. Composite Materials: Design and Application. Boca Raton: CRC Press LLC; 2003. 523 p. ISBN 1587160846

[11] Moksha O. SibNIA Completes Research on the CTP-40DT. [Internet]. Aeroo.ru Aviation News. 2019. Available from: https://aeroo.ru/30135-SibNIA-zavershil-raboty-po-samolietu-STR-40DT.html [Accessed: November 28, 2023]

[12] Randall R, Solo L, Chen C. Unmanned Aerial Vehicle. Patent, China, ZL201621265147.4, 2017

[13] Chen C, Bo L. Unmanned Aerial Vehicle. Patent, China, ZL201621265305.6, 2017

[14] Randall R, Chen C. Aircraft and Aircraft Control System, USA, US 2019/0263513 Al, 2019

[15] Kretov A, Tiniakov D. Evaluation of the Mass and Aerodynamic Efficiency of a High Aspect Ratio Wing for Prospective Passenger Aircraft, Aerospace. MDPI; 2022;**9**(9):497. DOI: 10.3390/aerospace9090497

[16] Kretov A, Tiniakov D. Evaluation of wing structures at the conceptual stage of transport category aircraft projects. Aviation, Issue. 2022;**26**(4):235-243. DOI: 10.3846/aviation.2022.18041

[17] Kretov AS, Tiniakov DV, Shataev PA. Conceptual assessment of the fuel efficiency of passenger aircraft with the transition to composite wings. Civil Aviation High Technologies;**26**(2): 72-90. DOI: 10.26467/2079-0619-2023-26-2-72-90

[18] Tinyakov DV. Analysis of geometric parameters features of the wings of

modern transport category airplanes from the top-view. Open Information and Computer Integrated Technologies. 2013;**62**:14-20

[19] McKeegan N. 787 Dreamliner completes wing-bending test. [Internet]. New Atlas. 2010. Available from: https://newatlas.com/787-dreamliner-completes-ultimate-load-test/14663/?0000016e-d4b4-d036-a77f-defe204f0000-page=2 [Accessed: November 28, 2023]

[20] Turbli. How Much Can an Airplane Wing Bend? [Internet]. Turbli. 2022. Available from: https://turbli.com/blog/how-much-can-an-airplane-wing-bend/ [Accessed: November 28, 2023]

[21] Brushgens GS. Aerodynamic and dynamic of long-haul aircraft. In: Proceedings of the TsAGI and Avia PRC, Moscow-Beijing. Moscow-Beijing: TsAGI and Avia PRC; 1995. p. 772. ISBN 7800467899

[22] Trento C. Silicon Carbide Fibers Used in Composite Material Reinforcement [Internet]. Stanford Advanced Materials. 2022. Available from: https://www.samaterials.com/content/silicon-carbide-fibers-used-in-composite-material-reinforcement.html [Accessed: November 28, 2023]

[23] Tsai SW, Wu EM. A general theory of strength for anisotropic material. Journal of Composite Materials. 1971;**5**: 58-80

[24] Hashin Z. Failure criteria for unidirectional fiber composites. Journal of Applied Mechanics. 1980;**47**(2): 329-334. DOI: 10.1115/1.3153664

[25] Puck A, Schurmann H. Failure analysis of FRP laminates by means of physically based phenomenological models. Composites Science and Technology. 1988;**58**:1045-1067. DOI: 10.1016/s0266-3538(96)00140-6

[26] Tiniakov D, Li L, Su Y. Airworthiness Design of Civil Aircraft Systems and Structure. Beijing: Beijing Press; 2018. 366 p. ISBN 9787512427624

[27] Karpov Y. Design of Parts and Units from Composites. Kharkov: KhAI; 2010. 767 p

[28] Grishin V, Dzyuba A, Dudarkov Y. Strength and Stability of Elements and Joints in Composite Aircraft Structures. Moscow: TsAGI; 2013. p. 139. ISBN 9785940522287

[29] Golovan V, Dudarkov Y, Levchenko E, Limonin M. Load bearing capacity of composite panels with in-service damages. Journal Trudy MAI. 2020;**110**:26. DOI: 10.34759/trd-2020-110-5

[30] Dudarkov Y, Levchenko E, Limonin M, Shevchenko A. Computational studies of some types of operational and technological damages impact on bearing capacity of stringer panels made of composite fiber reinforced plastic. Journal Trudy MAI. 2019;**106**:24. DOI: 10.34759/trd-2020-110-5

Chapter 4

# Mechanical Behavior of High-Strength Steel Fiber-Reinforced Concrete

*Álvaro Mena-Alonso, Dorys C. González, Jesús Mínguez and Miguel A. Vicente*

## Abstract

Steel fiber-reinforced concrete (SFRC) is a composite material, consisting of conventional concrete with the addition of short, randomly distributed steel fibers. Fibers modify the mechanical behavior of concrete, improving some of its properties: they increase its ductility, enhance its residual tensile strength, and under certain conditions, increase its ultimate flexural strength. All these advantages make this material competitive with conventional reinforced concrete. However, the psychological barriers of the construction sector and the technical challenges that remain to be solved are slowing down the consolidation of this building material. One of these challenges is the improvement of the understanding of fatigue, which not only affects SFRC, but concrete in general. In this regard, work to date suggests that fibers, given certain circumstances, increase the fatigue life of concrete. This would result in SFRC being very effective in structures where fatigue is a critical action, such as wind turbine concrete towers.

**Keywords:** steel fiber-reinforced concrete, high-strength concrete, mechanical behavior, fatigue, flexural strength

## 1. Introduction

Fiber-reinforced concrete is a composite material used in construction. It is defined as a type of concrete that includes in its composition short, discrete, and randomly distributed fibers in its matrix (**Figure 1**) [1]. It should be noted that current standards only consider steel fiber-reinforced concrete (SFRC) as a structural material [1, 2]. This is because the degree of development is much higher compared to other fibers, such as polypropylene, carbon, or glass.

The function of fibers is to make up for the shortcomings of conventional concrete, such as brittleness and very low tensile strength. In this sense, fibers increase the capacity of concrete to absorb deformation energy, making it much more ductile. Additionally, they enhance the residual tensile strength, or even the ultimate tensile strength [3, 4]. This leads to multiple benefits from the point of view of its structural functionality.

**Figure 1.**
*Cracking surface of a steel fiber-reinforced concrete specimen.*

The use of fibers to improve the properties of building materials is an ancestral technique, which originated more than 10,000 years ago in adobe blocks, made of mud reinforced with straw. In the case of concrete, the addition of fibers began to be applied at the beginning of the 20th century, practically after the emergence of modern concrete made with Portland cement. The first patents using steel fibers to improve the mechanical response of concrete were those of Graham (1911, USA) and Alfsen (1918, France). During the following decades, patents continued to be developed, using mainly steel and glass fibers.

In the 1950s and 1960s, the research and development of fiber-reinforced concrete gained momentum. The efforts of researchers focused on perfecting the geometry of the fibers to solve the problems of workability and lack of adhesion to the concrete matrix, which were very common at this stage. In this regard, the work of Romualdi in the mid-1960s stands out [5, 6]. At this time, steel fiber-reinforced concrete began to be commonly used in applications such as high-performance pavements, foundations for heavy machinery, and air-raid shelters.

During the following decades and up to the present, very significant advances have been made in the performance of SFRC. One of them is the fibers with shaped ends (**Figure 2**), which improve adhesion with respect to straight fibers, increasing the ductility of the material [7, 8]. Another is gluing (**Figure 2**), which reduces the risk of fiber balling during batching, facilitating the use of very slender fibers, which are much more structurally efficient.

Not only the improvement of steel fibers has contributed to the development of SFRC, but also the appearance of additives, among which superplasticizers, stands out. These admixtures make it possible to compensate for the worsening of consistency and workability caused by fibers without increasing the water/cement ratio and therefore without impairing the mechanical response of the concrete [9, 10].

This set of advances has made it possible to broaden the field of application of SFRC, which extends from the construction of pavements to withstand heavy loads and impacts, to the lining of subway works in the form of fiber-reinforced shotcrete. However, there are still significant technical and psychological barriers to giving fibers a greater structural responsibility, so that they partially or totally replace

**Figure 2.**
*Glued steel fibers with shaped ends.*

conventional rebar reinforcement. Therefore, the scientific community continues to move forward with the common goal of establishing SFRC as a structural material.

This chapter presents the latest scientific advances in the study of the mechanical behavior of steel fiber-reinforced concrete. Section 2 describes the components of SFRC, as well as some notions on dosage criteria. Section 3 describes the strength mechanism of SFRC. Section 4 then presents the mechanical properties of this material. Section 5 discusses the fatigue response of SFRC, a field of research that is currently attracting a lot of interest. Finally, Section 6 presents the conclusions.

## 2. Components and dosage

Steel fiber-reinforced concrete consists of a concrete matrix to which the fibers are added. The components of the matrix are the same as those of conventional concrete: cement, water, aggregates, additions, and admixtures. However, the design of SFRC presents certain differences, as the addition of fibers modifies the properties of concrete, affecting the requirements of the rest of the components.

The most relevant aspects of each component of SFRC are described below, as well as the most common dosage criteria.

### 2.1 Cement

The type of cement used in the manufacture of SFRC is the same as that used in conventional concretes. When high strength is sought, cement with a high strength class is used, such as Portland cement CEM I 52.5 R [11].

As for the cement content, it is usually high, with minimum dosages of about 400 kg/m$^3$. This is an essential requirement to achieve high strength, but there are also other reasons for this. Increasing the volume of cement paste improves the workability of concrete, counteracting the negative effect of the fibers. On the other hand, it is becoming increasingly common to work with self-compacting SFRC, which has multiple advantages, such as better workability or the absence of vibration. This type of concrete requires a large amount of fines to maintain cohesion and avoid segregation problems.

### 2.2 Water

There are no special requirements for the water used in the manufacture of SFRC, other than the usual ones, such as avoiding the use of types of water likely to promote fiber corrosion and chemical attack on the concrete matrix.

The water/cement (w/c) ratio is low, in the order of 0.4. This is a condition for obtaining high strength. If the w/c ratio was high, a significant volume of water would not react with the cement, causing an increase in porosity and thus a decrease in strength.

### 2.3 Aggregates

The aggregates that constitute concrete can be classified into three large groups according to their size: coarse aggregate or gravel (>2 mm), fine aggregate or sand (<2 mm and > 0.063 mm), and fines (<0.063 mm). The design of a suitable particle size is essential in the manufacture of SFRC, as it directly affects consistency, workability, and strength.

As for the coarse aggregate, it must be small enough to allow adequate fiber distribution (**Figure 3**). For this reason, it is recommended that the ratio between coarse aggregate size and fiber length be a maximum of 0.5 [1]. With respect to the fine aggregate, its quantity is usually high, with fine aggregate/coarse aggregate ratios greater than 1. In this way, segregation problems are avoided and fiber distribution in the matrix is favored. Finally, a high fines content is very important, as it improves workability and increases cohesion. The minimum fines content should be about 500 kg/m$^3$, which can be added in the form of filler or by means of pozzolanic additions, such as silica fume.

### 2.4 Admixtures

As mentioned above, fibers reduce the fluidity of concrete, worsening its workability. To solve this problem, water can be added, but this would lead to a decrease in concrete strength. In this context, the use of superplasticizing admixtures is highly recommended. These admixtures improve the consistency of fresh concrete without increasing the w/c ratio.

The dosage of the superplasticizing admixture will be indicated by the manufacturer, and it is not usual to exceed 4% by the weight of cement. It should be

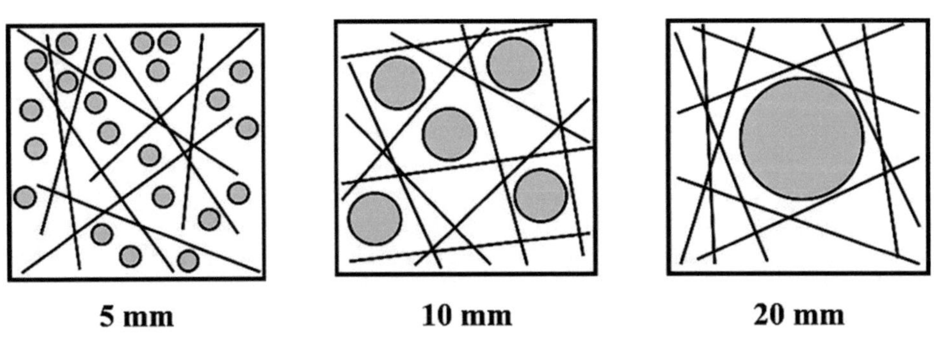


**Figure 3.**
*Effect of aggregate size on fiber distribution [12].*

mentioned that an excessive amount of superplasticizer may result in problems of segregation or exudation of the cement paste.

## 2.5 Fibers

Fibers are the differentiating component of SFRC with respect to conventional concrete and are what give it the status of a composite material. They are linear elements of short length and small sections that are added to the concrete matrix to improve some of its properties. In addition to steel fibers, discussed here, there are many other types: polypropylene, glass, carbon, natural, and so on.

Fibers can have different shapes, such as straight, wavy, serrated, crimped, with hooked ends or with flat ends (**Figure 4**). The most common ones are straight and hooked-end fibers. It is important to emphasize that fiber geometry significantly influences fiber-matrix adhesion and thus the mechanical response of SFRC. Higher adhesion increases the energy absorption capacity of the material and therefore its ductility.

The main geometrical properties of fibers are length and slenderness. On the one hand, certain criteria have to be considered when choosing the fiber length. A maximum fiber length to minimum mold or formwork dimension ratio of 1/3 is recommended [13, 14]. This reduces the wall effect, which causes the fibers closest to the mold to tend to be oriented parallel to the mold faces. The control of this phenomenon is very important, especially when extrapolating laboratory results, obtained on small samples, to full-scale structures.

On the other hand, fiber slenderness, defined as the ratio between length and diameter, is related to their mechanical efficiency. For the same volume of fibers, the greater the slenderness, the greater the adhesion to the matrix and the ductility of the

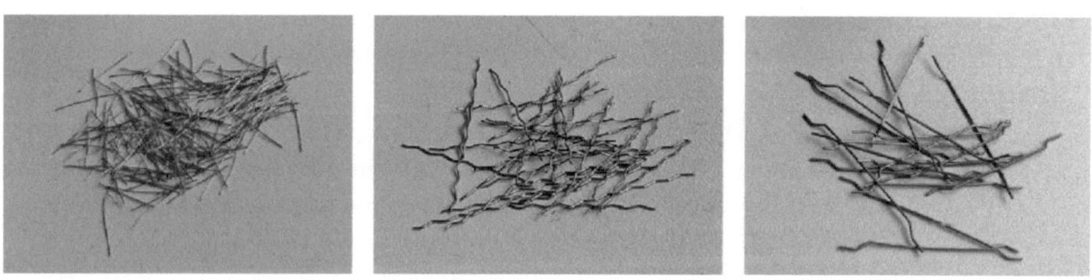

**Figure 4.**
*Types of steel fibers. From left to right, straight, crimped, and hooked-end.*

| | | |
|---|---|---|
| **Geometrical properties** | **Length (mm)** | **10–60** |
| | Equivalent diameter (mm) | 0.2–1.0 |
| | **Slenderness** | **45–80** |
| Material and mechanical properties | Modulus of elasticity (GPa) | >200 |
| | Tensile strength (MPa) | 500–3000 |
| | Elongation at break (%) | 2–7 |
| | Density (kN/m$^3$) | 78.5 |

**Table 1.**
*Properties of steel fibers for SFRC.*

SFRC. However, very high slenderness (>100) considerably reduces the workability of concrete and favor the formation of fiber balls.

**Table 1** shows the average values of the main properties of steel fibers.

Fiber content is expressed as the weight of fibers per unit volume of concrete ($kg/m^3$) or as the percentage of fiber volume in relation to the volume of concrete. The minimum dosage of fibers when they fulfill a structural function is of the order of 20 $kg/m^3$, while the maximum, limited by the decrease in workability, is in the range of 160 $kg/m^3$. In any case, the most common range of fiber contents is between 25 and 95 $kg/m^3$.

## 3. Strength mechanism

The main function of the fibers is to withstand tensile stresses because, as mentioned above, the tensile strength of plain concrete (PC) is low.

Consider an SFRC beam subjected to a three-point bending test (**Figure 5**). As the load increases, so does the maximum tensile stress. As long as the stress is low, the tensile stress will be supported by the concrete matrix, with little involvement of the fibers. This will continue until concrete cracks, i.e., until the stress corresponding to the first crack strength is reached, which is approximately around 3–5 MPa. At this point, a process of progressive transfer of tensile stresses from the matrix to the fibers will begin. Finally, when concrete is in a fully cracked state, the fibers will be the only ones responsible for supporting the tensile stresses.

Steel fibers have two main failure mechanisms: pull-out, due to the exhaustion of the tangential bond stresses with the matrix ($\tau > \tau_{max}$) and breakage, due to the exhaustion of the normal tensile stresses ($\sigma > \sigma_{max}$). In practice, fiber failure in SFRC is mostly due to pull-out, i.e., most commonly, the ultimate tangential pull-out bond stress $\tau_{max}$ is reached earlier than the ultimate tensile normal stress $\sigma_{max}$. This is positive from the point of view of structural safety as the pull-out failure mode is more progressive and ductile.

The strength of steel fibers against pull-out depends significantly on fiber geometry. The two most common types will be analyzed: straight and hooked-end. In addition, it will be considered that fibers have a certain inclination with respect to the crack plane, as, being randomly distributed, it is not expected that they are perfectly orthogonal to it.

When a straight fiber is subjected to tension, the acting mechanisms are adhesion, friction, and mechanical anchorage due to inclination (**Figure 6**) [15]. Adhesion is the

**Figure 5.**
*Three-point bending test on the SFRC specimen. The final phase in fully cracked state.*

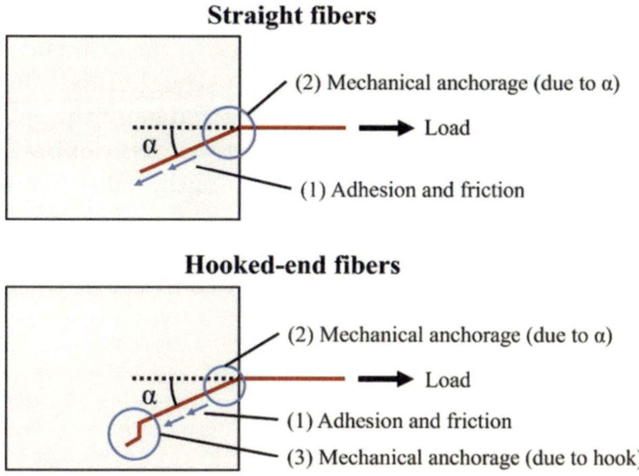


**Figure 6.**
*Pull-out strength mechanisms in straight and hooked-end fibers.*

force that binds the fiber to the concrete matrix and depends on the microstructure of the fiber-matrix interface. Friction is the force that is mobilized when the fiber is partially or completely detached from the matrix. Finally, when the fiber is tilted, the force acting in the direction of its axis is a part of the total pull-out force, thus generating a mechanical anchoring effect.

It is observed that mechanical anchorage has a positive impact on pull-out behavior, that is, fibers that have a certain inclination with respect to the crack plane have more pull-out strength than those that are perfectly orthogonal to it (**Figure 7**). The pull-out strength increases with the inclination of the fibers up to a certain angle, after which it decreases drastically, as the failure occurs by the detachment of the concrete cover. According to several studies, the optimum angle of inclination for which the maximum pull-out strength occurs ranges from 30° to 60°, although it depends on fiber geometry and matrix properties [16, 17].

On the other hand, in hooked-end fibers, the mechanisms that act when subjected to tensile stress are the same as those of straight fibers and in addition the mechanical anchorage by the hooked end (**Figure 6**). In this case, the impact of the anchorage

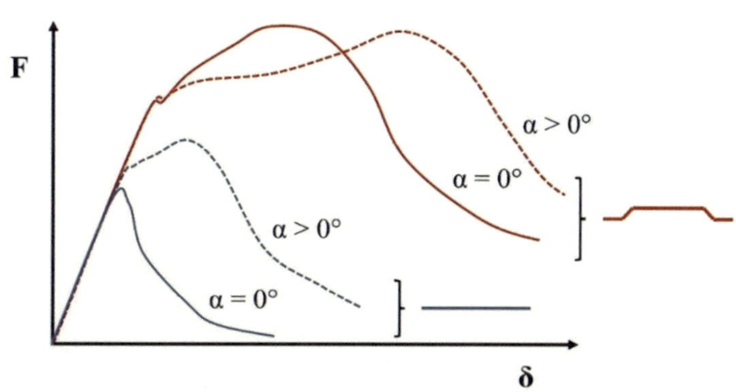


**Figure 7.**
*Approximate force vs. displacement diagrams in pull-out tests of straight and hooked-end fibers with and without inclination.*

due to the hook is very significant, with two fundamental results. First, the pull-out strength is higher than that of straight fibers. Secondly, the influence of the inclination on the pull-out strength is much lower than for straight fibers. In fact, for moderate inclinations (up to 30°), and depending on fiber and matrix properties, pull-out strength remains constant [16, 18]. However, it is worth mentioning that, although the inclination does not necessarily increase the strength, it does increase the energy absorbed by the fibers (**Figure** 7).

From the above, it is concluded that hooked-end fibers are more efficient than straight fibers, as the additional mechanism generated by the hook increases both the pull-out strength and the ductility.

## 4. Mechanical properties

Some of the characteristics of plain concrete are high compressive strength, very low tensile strength, and brittleness at breakage. The presence of fibers causes these properties to change to a greater or lesser extent. The main mechanical properties of SFRC are described below.

### 4.1 Compressive strength

Numerous authors conclude that the addition of fibers, as well as an increase in their content, do not necessarily improve the compressive strength of concrete [19–21]. This is explained by the balance between two opposing effects: on the one hand, the positive effect of the bridging forces generated by the fibers, which reduces the occurrence and growth of cracks and on the other hand, the negative effect of the reduction of fluidity that they induce, increasing porosity.

In any case, fibers do affect the stress-strain diagram ($\sigma$-$\varepsilon$). It is observed that the strain under maximum load is higher in SFRC than in plain concrete, which is explained by the ductility they provide. In addition, SFRC may have a certain residual compressive strength capacity, which is not observed in PC.

### 4.2 Modulus of elasticity

It is observed that the usual fiber contents do not produce relevant improvements of the modulus of elasticity, so it can be assumed that it remains constant [21].

This is due to the fact that the modulus of elasticity characterizes the linear-elastic behavior of concrete. At this stage, it can be considered that concrete is intact and has not yet cracked. Therefore, due to strain compatibility, the matrix resists practically the entire load, while the fibers have hardly been mobilized. Consequently, the stiffness of the SFRC in this range is the stiffness of the matrix, i.e., the same as that of plain concrete.

### 4.3 Tensile strength

In general, the ultimate direct tensile strength of SFRC is not higher than that of PC. However, when certain conditions are met, an increase in tensile strength is observed. This is the case for ultra-high-performance fiber-reinforced concrete, characterized by high energy absorption capacity and strain hardening behavior [4]. Some of the factors contributing to achieve these properties are high fiber content

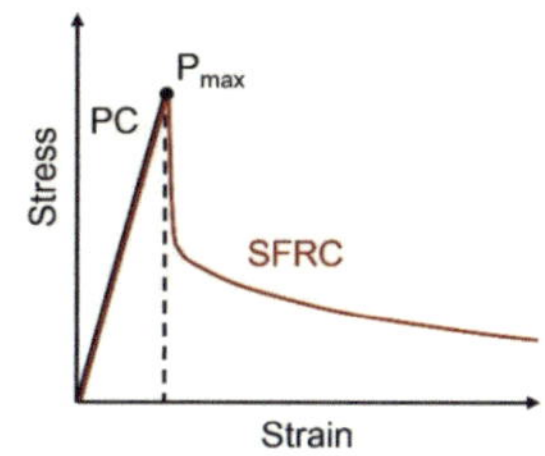


**Figure 8.**
*Stress–strain diagram in plain concrete and fiber-reinforced concrete under uniaxial tension.*

(≥1% or 78.5 kg/m$^3$) and to a lesser extent high slenderness (≥70), efficient fiber geometry (e.g., hooked-end fibers), and good fiber-matrix adhesion.

Although, in conventional SFRC, the addition of fibers does not lead to an increase in direct ultimate tensile strength, a significant increase in residual tensile strength is observed. As **Figure 8** shows, in the σ-ε diagram of PC, once the maximum stress is reached, there is a sharp drop due to highly brittle failure. However, in SFRC, after reaching the maximum stress, the drop is not so pronounced and fibers are able to withstand a certain level of stress up to high deformations. This residual strength contribution is known as strain softening.

### 4.4 Flexural strength

The flexural behavior of SFRC is very different from that of PC. In all cases, a very significant increase in residual flexural strength is observed. As for the maximum flexural strength, its value can be maintained or increased, which depends on a set of factors: fiber volume, slenderness, typology, fiber-matrix adhesion, and orientation, among others. However, it should be noted that fibers have a much greater impact on the flexural response of concrete than on the direct tensile response [4]. In other words, the requirements of SFRC components to achieve flexural strength greater than that of plain concrete are lower than in the case of uniaxial tension. For example, in some works, an increase in flexural strength is observed with relatively low fiber contents (0.6%) [21, 22].

Standardized tests to evaluate the flexural strength of SFRC consist of prismatic specimens, with or without notching, subjected to bending at three or

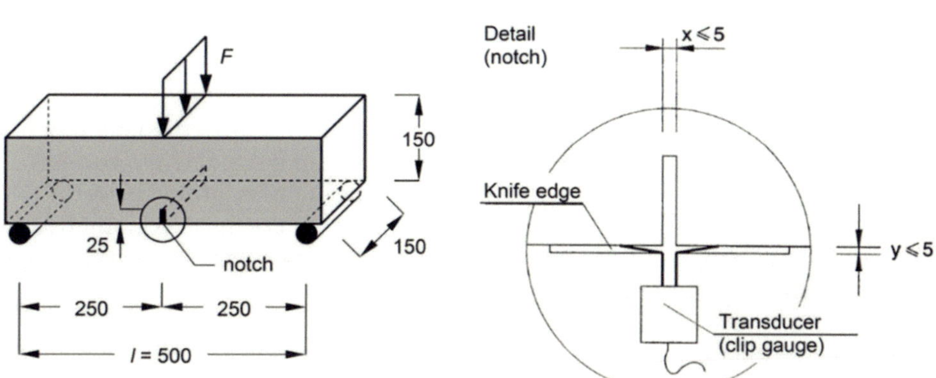


**Figure 9.**
*SFRC flexural test configuration according to [14].*

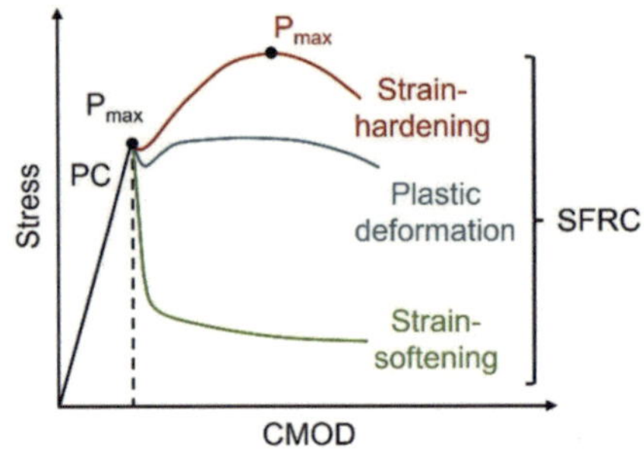


**Figure 10.**
*Stress-CMOD (crack mouth opening displacement) diagram of flexural test on plain concrete and steel fiber-reinforced concrete with different behaviors.*

four points. The most common is the one described in EN 14651 (**Figure 9**) [14], equivalent to RILEM TC 162-TDF [23]. It is a three-point bending test performed on 150 × 150 × 600 mm prismatic specimens with a 25 mm deep central notch.

The main result of a flexural test in SFRC is the stress-crack opening (CMOD) diagram. **Figure 10** compares this diagram in PC and various types of SFRC.

**Figure 10** shows that the first section with linear-elastic behavior is the same in all cases. This is because concrete has not yet cracked, and therefore, the fibers have not been mobilized. From this, it can be concluded that the first crack strength ($P_f$) depends exclusively on the matrix, at least for normal fiber contents.

When concrete cracks, brittle fracture of the specimen occurs in PC. However, in SFRC, three situations can occur:

a. Strain-softening: the bridging forces of the fibers do not withstand more stress than the matrix, so the stress decreases with crack opening. However, a certain residual stress and thus a significant increase in strain energy absorption capacity is observed. In this case, the ultimate flexural strength coincides with the first crack strength ($P_{max} = P_f$).

b. Plastic deformation: the strength mechanism of the fibers supports a stress similar to that of the matrix, so that the stress remains sensibly constant up to high values of CMOD. In this situation, it is observed that the maximum flexural strength is approximately equal to the first crack strength ($P_{max} = P_f$). However, the difference with respect to the previous case is that the residual strength is much higher.

c. Strain hardening: the bridging forces of the fibers are able to withstand more stress than the matrix, which increases the maximum flexural strength. This maximum load occurs for high CMODs due to the ductility provided by the fibers. This is the only case where $P_{max} > P_f$, that is, where the maximum flexural stress is greater than that of plain concrete. This increase varies depending on the characteristics of the SFRC and can be more than 100%.

These three situations can also be observed in direct tensile tests, although with much higher fiber contents. It is possible to have an SFRC that shows strain-softening under direct tension, but strain-hardening under bending.

### 4.5 Fatigue

As the fatigue behavior of SFRC is a field of research of great scientific interest at present, this issue will be addressed separately in Section 5.

### 4.6 Other mechanical properties

In addition to those already mentioned, the addition of steel fibers improves other concrete properties, such as impact strength and abrasion resistance. Additionally, fibers can increase the durability of concrete, as they control cracking, distributing it, and reducing the crack width. This hinders reinforcement corrosion and increases resistance to chloride attack, among others.

## 5. Fatigue behavior of SFRC

### 5.1 Definition and interest of fatigue

Fatigue is defined as a process of mechanical degradation of a material leading to its failure, which occurs when it is subjected to cyclic loading. This phenomenon is of great interest because it implies that a given element can collapse without ever having exceeded its strength under static conditions.

In the case of concrete, the study of fatigue is very interesting for several reasons. First, practically, all concrete structures are subjected to cyclic loading, to a greater or lesser extent. Thus, there are structures that undergo few load cycles, but with a very high stress level (such as a building subjected to an earthquake), while others support a large number of cycles with a low stress level (such as a dam subjected to waves). Secondly, fatigue is an increasingly relevant phenomenon, and all structural codes devote more space to it, as, with the advance of high-strength concrete, structures are becoming increasingly slender. As a result, the importance of variable loads (wind, traffic, etc.), which are cyclic in nature and therefore susceptible to fatigue, is growing. In fact, in structures such as wind turbine concrete towers, fatigue is the dimensioning action.

Thirdly, and related to the above, concrete fatigue has a wide dispersion of results, being common that the fatigue life of two identical elements subjected to the same load cycles can vary by up to two orders of magnitude. This lack of knowledge of the phenomenon leads structural codes to be very conservative in fatigue design. Finally, the field of flexural fatigue has been of great scientific interest in recent years, particularly in steel fiber-reinforced concrete. This is because the fibers are able not only to provide residual flexural strength but also to increase the ultimate flexural strength itself. It is expected that, in the coming years, structural solutions based on SFRC will be competitive with the current reinforced concrete solutions.

### 5.2 Impact of fibers on the fatigue response of concrete

The addition of fibers modifies the mechanical characteristics of concrete, and therefore its fatigue response. There are numerous works that study the influence of the presence and content of fibers on the fatigue life of concrete; that is, on the number of stress cycles between certain maximum and minimum levels that a given element can withstand until failure.

Germano et al. [24] carried out an experimental study of flexural fatigue in SFRC, considering different fiber contents (0%, 1%, and 2%) and load levels (15–65%, 25–75%, and 35–85% with respect to the static flexural strength). The results are very interesting (**Figure 11**). On the one hand, it is observed that the fatigue life of SFRC is higher than that of PC for high load levels; however, this positive effect decreases with decreasing load levels. On the other hand, in all cases, the fatigue life of specimens with 0.5% fibers is higher than those with 1% fibers. This suggests that there is an optimum fiber content in terms of fatigue response.

Other research also suggests that fibers improve the fatigue life of concrete, although only up to a given content [25–27]. Vicente et al. [26] performed flexural fatigue tests on precracked specimens with fiber contents of 1 and 2%. They concluded that, while the residual flexural strength increased with the number of fibers, the fatigue life remained practically constant. Poveda et al. [25] studied the influence of different fiber contents (from 0 to 0.8% by volume) on the compressive fatigue response of self-compacting concrete. They observed that the optimum fiber content, for which the fatigue life increased by up to five times compared to that of plain concrete, was 0.6%. Moreover, at 0.8% and above, there was a marked decrease in fatigue life, attributed to the increase in porosity induced by the high fiber content.

In summary, it can be concluded that the addition of fibers improves the fatigue response of concrete, both in bending and compression, but only under certain conditions of load levels and fiber contents. Therefore, further research in this field is required to obtain more robust results.

The factor related to the fibers that have been most studied with regard to fatigue strength is the content; however, there are others that may be relevant, such as fiber typology, slenderness, or orientation. In the case of orientation, most publications analyze its influence on the static response of concrete. It is shown that the residual flexural strength increases when the fibers are more perpendicular to the crack plane [28, 29]. With respect to fatigue, one of the few publications is that of Vicente et al. [30], which studies the influence of orientation on the compressive fatigue of SFRC.

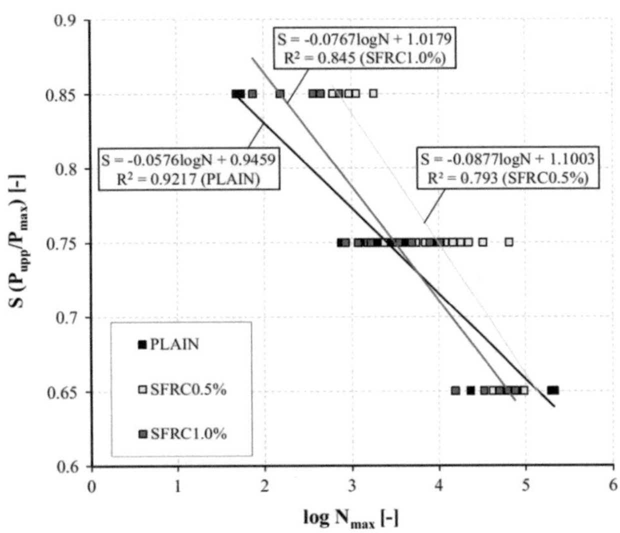


**Figure 11.**
*Stress-number of cycles (S-N) curves for plain concrete and SFRC with fiber contents of 1% and 2%, according to [24].*

By applying microcomputed tomography, they conclude that the fatigue life is higher when the fibers are arranged more perpendicular to the loading axis. This is explained by the bridging forces of the fibers, which prevent the propagation of vertical cracks.

### 5.3 Fatigue damage

Cyclic fatigue loading causes progressive damage to the concrete. This damage manifests itself internally with the appearance and growth of cracks. Fatigue cracks originate in the weakest zones of the SFRC, which are the coarse aggregate-matrix and fiber-matrix interfaces [30, 31] (**Figure 12**).

Microstructural fatigue damage is reflected at the macrostructural level by some indicators, allowing it to be analyzed quantitatively. In compressive fatigue, fatigue cycles cause changes in the stress-strain diagram. In particular, there is an increase in the residual strain and a reduction in the dynamic modulus of elasticity [32]. Equivalently, in flexural fatigue, variations in the load-CMOD diagram are observed (**Figure 13**): an increase in the plastic or residual crack opening ($CMOD_{pl}$) and a decrease in stiffness ($k_i$) defined as the slope of the loading branch of each cycle [24, 33, 34].

Another diagram that allows analysis of the mechanical degradation of concrete due to flexural fatigue is the cyclic creep curve. It is the upper envelope of the CMOD vs. number of cycles diagram. In the most general case of a flexural fatigue test in SFRC (with a configuration as shown in **Figure 5**), this diagram looks like the one in **Figure 14**.

**Figure 14** clearly distinguishes two mechanisms of fatigue damage. First, there is a first phase in which the concrete has not cracked and the loads are supported by the matrix, with little intervention of the fibers. This mechanism is known as "fatigue of uncracked concrete" (in green in **Figure 14**) and consists of the following stages: (I) plastic tensile deformation and birth of microcracks at the notch edge, (II) stable growth of the microcracks, and (III) convergence into a main macrocrack that triggers failure. Second, when this mechanism is exhausted and concrete cracks, "fatigue of cracked concrete" (in red in **Figure 14**) begins. The stages are (I) concrete cracking and fiber loading, (II) stable propagation of the main macrocrack due to progressive fiber failure, and (III) unstable propagation of the macrocrack until failure. Once this second mechanism is exhausted, the collapse of the element occurs.

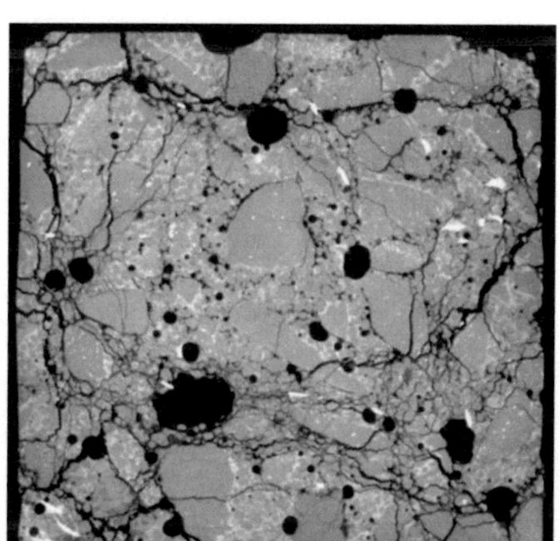

**Figure 12.**
*Microcomputed tomography image showing compressive fatigue damage in a cubic SFRC specimen [30].*

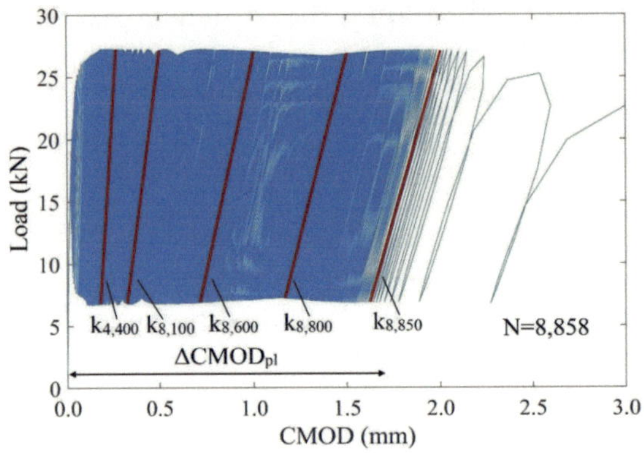


**Figure 13.**
*Load-CMOD diagram in a flexural fatigue test on SFRC [22].*

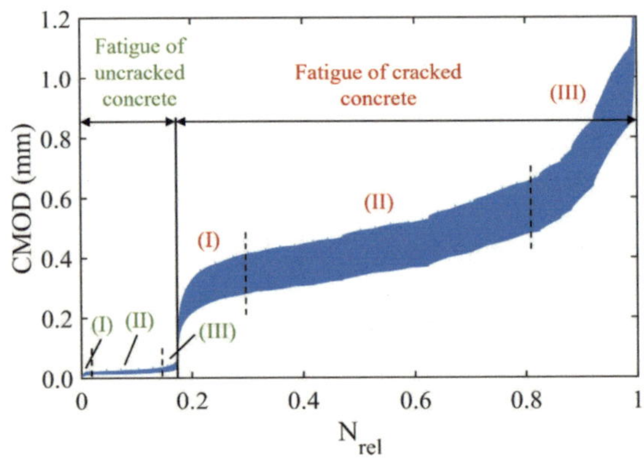


**Figure 14.**
*CMOD vs. relative number of cycles diagram in SFRC subjected to flexural fatigue [21].*

As can be seen, fatigue of cracked concrete is a much more ductile mechanism (higher CMOD variations) than fatigue of uncracked concrete. This is because the loads are supported by the bridging forces of the fibers, which are less stiff than those of the concrete matrix.

It is worth mentioning that, in SFRC, the fatigue of uncracked concrete does not always occur. This requires that the maximum fatigue stress level be lower than the first crack strength. Otherwise, the concrete would crack after the first loading cycle.

Finally, there is a parameter of the cyclic creep curves that is of great interest because of its relationship with fatigue strength. **Figure 14** shows that, in both fatigue mechanisms, stage (II) can be fitted to a straight line. The slope of this line is called the crack opening rate (dCMOD/dn). Several studies show that there is a clear correlation between the crack opening rate and fatigue life, such that the lower the dCMOD/dn, the higher the N [21, 24, 33]. This experimental relationship, known as the Sparks and Menzies relationship [35], was originally proposed in compressive fatigue. Its major utility is that it helps to understand the experimental dispersion of fatigue and can serve as a basis for the development of more robust fatigue design criteria.

## 6. Conclusions

Steel fiber-reinforced concrete is a composite building material that has undergone great development in terms of mechanical performance in recent years. However, the great inertia toward change in the construction sector, together with technical challenges still unresolved, meaning that SFRC has not yet been consolidated as a structural material of reference.

SFRC is basically a type of concrete to which short, discrete, and randomly distributed steel fibers are added. However, the addition of fibers affects the requirements of the other components. In general, a larger volume of cement, smaller coarse aggregate, and superplasticizing admixtures are needed to improve workability without compromising strength.

Fibers improve the fracture behavior of concrete, resulting in SFRC having a higher energy absorption capacity and thus higher ductility than plain concrete. In addition, the bridging forces of the fibers increase the tensile strength, which in PC is very low. In this sense, SFRC is particularly efficient in elements subjected to flexural stress.

The study of fatigue in SFRC is of great interest due to the possibilities it offers in structures where this action is dimensioning, such as wind turbine concrete towers. Several studies suggest that the addition of fibers increases the fatigue life of concrete, although only under certain load levels and fiber contents.

In conclusion, although there have been very relevant advances, there are still important technical challenges to be solved in order to optimize the mechanical behavior of steel fiber-reinforced concrete.

## Acknowledgements

The authors are grateful for the financial support of the Spanish Ministerio de Economía y Competitividad (grant no. PID2019-110928RB-C32) and the Spanish Ministerio de Ciencia, Innovación y Universidades (grant no. FPU19/02685).

## Author details

Álvaro Mena-Alonso*, Dorys C. González, Jesús Mínguez and Miguel A. Vicente
University of Burgos, Burgos, Spain

*Address all correspondence to: amena@ubu.es

## References

[1] European Committee for Standardisation (CEN), EN 1992-1-1:2023. Eurocode 2 - Design of Concrete Structures - Part 1-1: General Rules and Rules for Buildings, Bridges and Civil Engineering Structures. Brussels, Belgium. 2023

[2] International Federation for Structural Concrete (fib). Model Code 2010. Lausanne, Switzerland. 2012

[3] Choi SW, Choi J, Lee SC. Probabilistic analysis for strain-hardening behavior of high-performance fiber-reinforced concrete. Materials. 2019;**12**:2399. DOI: 10.3390/ma12152399

[4] Wille K, El-Tawil S, Naaman AE. Properties of strain hardening ultra high performance fiber reinforced concrete (UHP-FRC) under direct tensile loading. Cement and Concrete Composites. 2014;**48**:53-66. DOI: 10.1016/j.cemconcomp.2013.12.015

[5] Romualdi JP, Mandel JA. Tensile strength of concrete affected by uniformly distributed and closely spaced short lengths of wire reinforcement. ACI Journal Proceedings. 1964;**61**:657-672. DOI: 10.14359/7801

[6] Romualdi JP, Batson GB. Mechanics of crack arrest in concrete. Journal of the Engineering Mechanics Division. 1963;**89**:147-168. DOI: 10.1061/JMCEA3.0000381

[7] Marcalikova Z, Cajka R, Bilek V, Bujdos D, Sucharda O. Determination of mechanical characteristics for fiber-reinforced concrete with straight and hooked fibers. Crystals (Basel). 2020;**10**:1-21. DOI: 10.3390/cryst10060545

[8] Qiu M, Shao X, Zhu Y, Hussein HH, Li F, Li X. Effect of aspect ratios of hooked end and straight steel fibers on the tensile strength of UHPFRC. Journal of Materials in Civil Engineering. 2022;**34**(7). DOI: 10.1061/(asce)mt.1943-5533.0004283

[9] de la Rosa Á, Poveda E, Ruiz G, Cifuentes H. Proportioning of self-compacting steel-fiber reinforced concrete mixes based on target plastic viscosity and compressive strength: Mix-design procedure & experimental validation. Construction and Building Materials. 2018;**189**:409-419. DOI: 10.1016/j.conbuildmat.2018.09.006

[10] ACI Committee 544, ACI 544.3R-93. Guide for Specifying, Proportioning, Mixing, Placing, and Finishing Steel Fiber Reinforced Concrete. Farmington Hills, MI, USA. 1998

[11] European Committee for Standardisation (CEN), EN 197-1. Cement. Part 1: Composition, Specifications and Conformity Criteria for Common Cements. Brussels, Belgium. 2011

[12] Asociación Española de Hormigón Estructural. M-2. Manual de tecnología del hormigón reforzado con fibras de acero. Madrid, Spain. 2000

[13] ASTM C1609/C1609M-12. Standard Test Method for Flexural Performance of Fiber-Reinforced Concrete (Using Beam with Third-Point Loading). West Conshohocken, PA, USA. 2012

[14] European Committee for Standardisation (CEN), UNE-EN 14651. Test Method for Metallic Fibre Concrete. Measuring the Flexural Tensile Strength

(Limit of Proportionality (LOP), Residual). Brussels, Belgium. 2008

[15] Singh H. Steel Fiber Reinforced Concrete. Singapore: Springer Singapore; 2017. DOI: 10.1007/978-981-10-2507-5

[16] Song F. Steel fiber reinforced concrete under concentrated load [thesis]. Germany: Ruhr-Universität Bochum; 2017

[17] Poveda E, Yu RC, Tarifa M, Ruiz G, Cunha VMCF, Barros JAO. Rate effect in inclined fibre pull-out for smooth and hooked-end fibres: A numerical study. International Journal of Fracture. 2020;**223**:135-149. DOI: 10.1007/s10704-019-00404-7

[18] Gebuhr G, Pise M, Sarhil M, Anders S, Brands D, Schröder J. Analysis and evaluation of the pull-out behavior of hooked steel fibers embedded in high and ultra-high performance concrete for calibration of numerical models. Structural Concrete. 2019;**20**:1254-1264. DOI: 10.1002/suco.201900034

[19] Yang J, Chen B, Nuti C. Influence of steel fiber on compressive properties of ultra-high performance fiber-reinforced concrete. Construction and Building Materials. 2021;**302**(124104). DOI: 10.1016/j.conbuildmat.2021.124104

[20] Gesoglu M, Güneyisi E, Muhyaddin GF, Asaad DS. Strain hardening ultra-high performance fiber reinforced cementitious composites: Effect of fiber type and concentration. Composites Part B: Engineering. 2016;**103**:74-83. DOI: 10.1016/j.compositesb.2016.08.004

[21] Mena-Alonso Á, González DC, Mínguez J, Vicente MA. Size effect on the flexural fatigue behavior of high-strength plain and fiber-reinforced concrete. Construction and Building Materials. 2024;**411**:134424. DOI: 10.1016/j.conbuildmat.2023.134424

[22] Mena-Alonso Á. Flexural fatigue of high-strength plain and fiber-reinforced concrete: influence of mesostructure and study of size effect [thesis]. Spain: University of Burgos; 2023. DOI: 10.36443/10259/7855

[23] International Union of Laboratories and Experts in Construction Materials, Systems and Structures (RILEM), Recommendations of RILEM TC 162-TDF: Test and design methods for steel fibre reinforced concrete: Bending test. Paris, France. 2002

[24] Germano F, Tiberti G, Plizzari G. Post-peak fatigue performance of steel fiber reinforced concrete under flexure. Materials and Structures. 2016;**49**:4229-4245. DOI: 10.1617/s11527-015-0783-3

[25] Poveda E, Ruiz G, Cifuentes H, Yu RC, Zhang X. Influence of the fiber content on the compressive low-cycle fatigue behavior of self-compacting SFRC. International Journal of Fatigue. 2017;**101**:9-17. DOI: 10.1016/j.ijfatigue.2017.04.005

[26] Vicente MA, Mínguez J, González DC. Computed tomography scanning of the internal microstructure, crack mechanisms, and structural behavior of fiber-reinforced concrete under static and cyclic bending tests. International Journal of Fatigue. 2019;**121**:9-19. DOI: 10.1016/j.ijfatigue.2018.11.023

[27] Gao D, Gu Z, Zhu H, Huang Y. Fatigue behavior assessment for steel fiber reinforced concrete beams through experiment and fatigue prediction model. Structure. 2020;**27**:1105-1117. DOI: 10.1016/j.istruc.2020.07.028

[28] Mínguez J, González DC, Vicente MA. Fiber geometrical parameters of fiber-reinforced high strength concrete and their influence on the residual post-peak flexural tensile strength. Construction and Building Materials. 2018;**168**:906-922. DOI: 10.1016/j.conbuildmat.2018.02.095

[29] Mudadu A, Tiberti G, Germano F, Plizzari GA, Morbi A. The effect of fiber orientation on the post-cracking behavior of steel fiber reinforced concrete under bending and uniaxial tensile tests. Cement and Concrete Composites. 2018;**93**:274-288. DOI: 10.1016/j.cemconcomp.2018.07.012

[30] Vicente MA, Ruiz G, González DC, Mínguez J, Tarifa M, Zhang X. Effects of fiber orientation and content on the static and fatigue behavior of SFRC by using CT-scan technology. International Journal of Fatigue. 2019;**128**:105178. DOI: 10.1016/j.ijfatigue.2019.06.038

[31] Zhang J, Stang H, Li VC. Fatigue life prediction of fiber reinforced concrete under flexural load. International Journal of Fatigue. 1999;**21**:1033-1049. DOI: 10.1016/S0142-1123(99)00093-6

[32] Holmen JO. Fatigue of concrete by constant and variable amplitude loading [thesis]. Norway: Division of Concrete Structures, Norwegian Institute of Technology, University of Trondheim; 1979

[33] Carlesso DM, de la Fuente A, Cavalaro SHP. Fatigue of cracked high performance fiber reinforced concrete subjected to bending. Construction and Building Materials. 2019;**220**:444-455. DOI: 10.1016/j.conbuildmat.2019.06.038

[34] Gebuhr G, Pise M, Anders S, Brands D, Schröder J. Damage evolution of steel fibre-reinforced high-performance concrete in low-cycle flexural fatigue: Numerical modeling and experimental validation. Materials. 2022;**15**:1179. DOI: 10.3390/ma15031179

[35] Sparks PR, Menzies JB. The effect of rate of loading upon the static and fatigue strengths of plain concrete in compression. Magazine of Concrete Research. 1973;**25**:73-80. DOI: 10.1680/macr.1973.25.83.73

Chapter 5

# Evaluation of Waste Tag Pins as Fibers in Gypsum Plasters

*Ahmet Hayrullah Sevinç, Muhammed Yasin Durgun and Hayriye Hale Aygün*

## Abstract

This study deals with the usability of tag pins on gypsum-based products, which are used to attach tags on goods such as textile products. The primary motivation of the study is that the tag pins become waste after the sale of the product and this waste is generally produced from polypropylene (PP), which is also used in fiber production. The study used waste tag pins in three different lengths (0.5, 1.0, and 1.5 mm) and at three different fiber volumes such as 5, 10, and 15%. Thus, 40 × 40 × 160 mm sized prismatic gypsum samples were produced and unit weight, ultrasonic pulse velocity (UPV), thermal conductivity, apparent porosity, water absorption, capillary water absorption, compressive and flexural strength of samples were tested. Samples with fiber content exhibited higher flexural strength than the reference sample. The use of longer fibers increased the flexural strength. As a result, the use of tag pins in gypsum matrix generally improved the pore structure and slightly increased the unit weight while enhancing properties such as porosity and water absorption. The same improvement was valid for the mechanical properties. However, the thermal insulation properties of gypsum-based products were adversely affected.

**Keywords:** tag pin, fiber, gypsum plaster, physical properties, mechanical properties

## 1. Introduction

Gypsum is a lightweight, easily processable, fast-setting, and environmentally friendly building material. Despite these advantages, it merges into the background with respect to cement mortar due to its poor mechanical characteristics, high shrinkage during heating, and low water resistance [1–3]. To overcome these shortcomings, gypsum is reinforced with various fibrous materials, such as polyester fiber [4], carbon fiber [5], glass fiber [6], sisal fiber [7], and plant-based fibers [8, 9], and has been widely used as gypsum-based composites in interior linings, such as wall panels, false ceiling, and partitions. In fiber-reinforced gypsum composites (FRGCs), the addition of fiber leads to improvement in the brittle characteristic of gypsum by increasing fracture toughness and crack propagation resistance [10].

However, the degree of improvement depends on fiber adhesion into the gypsum matrix and fiber type used for reinforcing gypsum to manufacture gypsum-based composites [11]. Many researchers have examined the effect of fiber type and fiber concentration on gypsum composite's mechanical resistance, water absorption, and thermal properties [12–17].

Tag pins, one of the mostly used auxiliary materials, have been commercially used in the textile industry for hooking price tags and labeling garments. They are generally manufactured from synthetic fibers, such as polyamide (PA), polypropylene (PP), thermoplastic urethane (TPU), and polyetheretherketone (PEEK). Still, PP tag pins are the mostly used type due to their low cost, easy processability, and producibility of various lengths with different colors. The use of PP tag pins is not only observable in the textile industry. They have also been used in many products to which price tags are attached, such as toys, accessories, household articles, footwear, etc. Considering that the increase in global textile waste is approximately 60% in each year, 57 million tons of waste is added onto deposited wastes in a year [18, 19]. It is reported that 2% wastage is available during the application of accessories to garments [20]. When its wide use is considered in many sectors, the total wastage during the fixing of pins onto garments and the consumption of tag pins is two pieces even for only a woven shirt, a considerable amount of tag pins are wasted since manufacturing and after a product is purchased by a consumer. There are a few research works on the evaluation and recycling of tag pin wastes for investigating the effect of these wastes on shallow slope failure [21–23].

Some researchers aimed to reinforce gypsum with PP fiber to produce a more versatile gypsum-based composite. Gencel et al. manufactured vermiculite/diatomite and PP (0.5–1%) fiber-reinforced gypsum composites. They reported that increasing PP fiber content caused an increase in sound velocity and a decrease in thermal diffusivity but did not affect the water absorption characteristic of the composite [24]. In another study, researchers pointed out that PP fibers improved the durability of gypsum composite better than that of glass or mineral fibers due to the good adhesion of PP fibers with gypsum matrix and large pores occurred by increasing PP content in gypsum composite [17]. Mohendesi et al. proposed analytical modeling for PP and poly-paraphenylene terephthalamide (PPTA) short-fiber reinforced gypsum composites. They concluded that PP and PPTA reinforcement resulted in a notable improvement in the tensile strength of the gypsum composite. Still, the impregnation of PPTA fiber into the gypsum matrix was more versatile than that of PP with a smooth surface [25]. Some researchers focused on determining the mechanical properties of PP-reinforced gypsum composites, resulting in an increase in flexural [26] and impact [17] strength, and gypsum water resistance was also improved [27]. Wang et al. examined the thermal properties of polyvinyl alcohol (PVA) and PP-reinforced gypsum-based foam insulation composite. They concluded that the degree of thermal conductivity was mainly related to the foam matrix, and fiber loading had no significant effect on thermal insulation [28].

This study differentiates from its counterparts by reinforcing the gypsum matrix with a waste PP tag pin. As mentioned above, there are only a few attempts to recycle PP tag pin waste. In this study, PP tag pins with a 0.5 mm fiber diameter were cut into various lengths and mixed with gypsum plaster by altering fiber loading. All samples were manufactured by following the same procedure and exposed to testing in the same conditions for a meaningful conclusion.

## 2. Materials and methods

### 2.1 Materials

In this study, plasterboard plaster, a commercial product produced in accordance with the TS EN 13279 standard, was used. The technical properties of the gypsum plaster used, which are stated by the company, are given in **Table 1**.

The images of various sizes of tag pins used in the study and the SEM image of a tag pin are shown in **Figure 1**. The tag pins were cut with scissors and brought to the desired dimensions. The technical properties of the tag pin fiber are given in **Table 2**. In the study, tap water was used as mixing water.

### 2.2 Methods

The mixture design (in g) and sample codes of the produced gypsum plaster samples are given in **Table 3**.

| Property | Value |
|---|---|
| Setting time—initial (min) | >8 |
| Setting time—final (min) | ≈30 |
| Min. compressive strength (40 × 40 mm) (MPa) | 10 |
| Min. flexural strength (40 × 40 × 160 mm) (MPa) | 4.5 |
| % retained by 200-μm sieve | 99.5 |
| % retained by 100-μm sieve | 95.0 |
| Loose unit weight (g/cm$^3$) | 0.75–0.80 |
| Dry unit weight (g/cm$^3$) | 1.05–1.10 |
| Absolute volumetric mass (g/cm$^3$) | 1.254 |

**Table 1.**
*Technical properties of the gypsum plaster.*

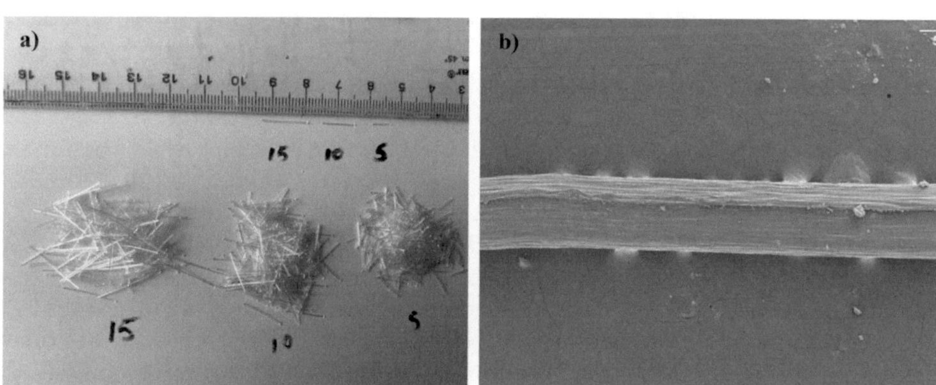


**Figure 1.**
*(a) Various sizes of tag pin fibers and (b) scanning electron microcopy (SEM) image of a tag pin fiber.*

| Property | Value |
|---|---|
| Fiber count (tex) | 136 |
| Fiber thickness (mm) | 0.5 |
| Tensile strength (MPa) | 135.93 ± 5.7 |
| Elongation (%) | 18.1 ± 0.6 |
| Density | 0.99 |
| Composition | 100% raw polypropylene |

**Table 2.**
*Technical properties of the tag pin fiber.*

| Code | Gypsum | Water | Fiber |
|---|---|---|---|
| R | 1320 | 858 | — |
| P5-0.5 | 1309.4 | 851.1 | 4.5 |
| P5-1 | 1298.8 | 844.2 | 9 |
| P5-1.5 | 1288.3 | 837.4 | 13.5 |
| P10-0.5 | 1309.4 | 851.1 | 4.5 |
| P10-1 | 1298.8 | 844.2 | 9 |
| P10-1.5 | 1288.3 | 837.4 | 13.5 |
| P15-0.5 | 1309.4 | 851.1 | 4.5 |
| P15-1 | 1298.8 | 844.2 | 9 |
| P15-1.5 | 1288.3 | 837.4 | 13.5 |

**Table 3.**
*Mixture design (in g) and sample codes of the produced gyssum plaster samples.*

The control mixture without tag pin fiber was named R, while the mixture with 10 mm length and 0.5% fiber was named P10-0.5. The mixture with a length of 5 mm and containing 1.5% fiber was expressed as P5-1.5. Tag pin fiber was used at the ratios of 0.5, 1, and 1.5% by volume, respectively. During the production, first of all, gypsum powder and fibers were dry-mixed for 60 s. When a completely homogeneous dry mixture was obtained, water was added. The mixtures were mixed in the laboratory-type cement mixer for 90 s until they became homogeneous. Then some of the produced mixture was placed in 40 mm × 40 mm × 160 mm prismatic molds and the rest in 120 mm × 120 mm × 20 mm plaque molds. Vibration was applied to the mixtures placed in the mold by lightly tapping the molds on the ground. Afterward, the finished samples were kept at 65% humidity and 21 ± 2° C temperature for 24 h to be set. After the samples removed from the mold were kept in the laboratory for 7 days, all samples were dried in the oven at 60° C for 48 h, and then the experiments whose test list and standards are given in **Table 4** were applied. Also SEM analysis was performed to examine the microstructure of the samples. **Figure 2** includes various images from the experimental study.

| Test | Standard |
|---|---|
| Unit weight | ASTM C 138 |
| Ultrasonic pulse velocity | ASTM C 597 |
| Water absorption | ASTM C 20 |
| Apparent porosity | ASTM C 20 |
| Capillary water absorption | TS EN 480-5 |
| Bending strength | TS EN 196-1 |
| Compressive strength | TS EN 196-1 |
| Thermal conductivity | ASTM C 1113 |

**Table 4.**
*Tests performed and the standards.*

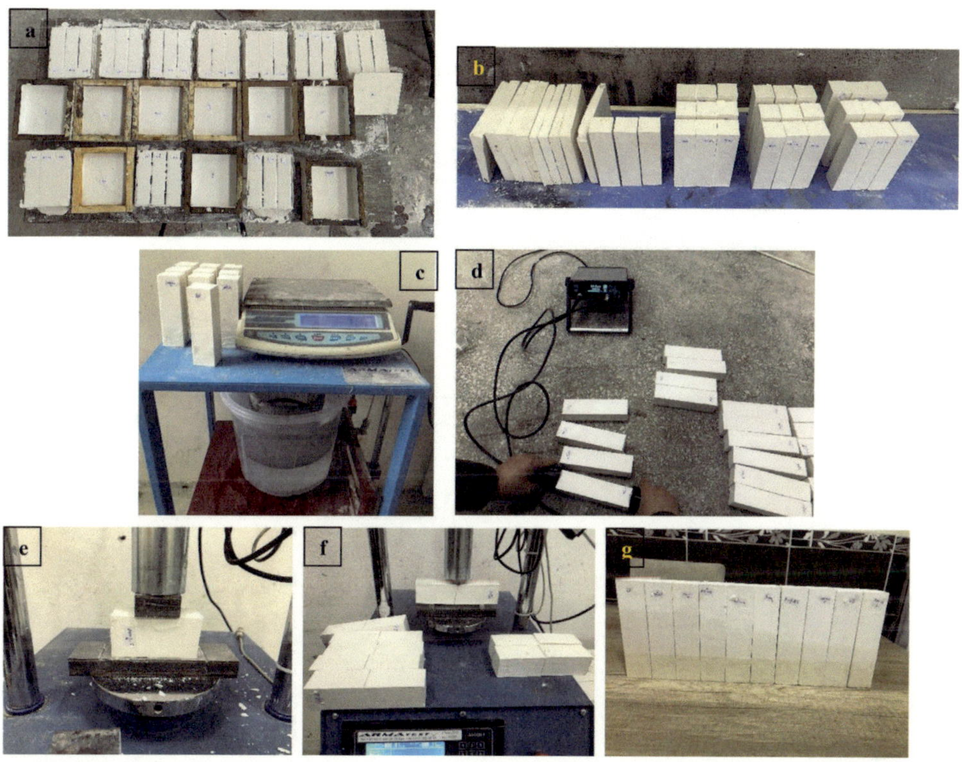


**Figure 2.**
*Images of experimental procedure. (a) Samples in molds, (b) all samples, (c) apparent porosity test, (d) ultrasonic pulse velocity test, (e) compressive strength test, (f) bending strength test, (g) capillary water absorbed samples.*

## 3. Results and discussion

The results of unit weight of samples are given in **Figure 3**. Unit volume weights of the samples changed between 1138 and 1206 kg/m$^3$. While the P15-1.5 sample gave

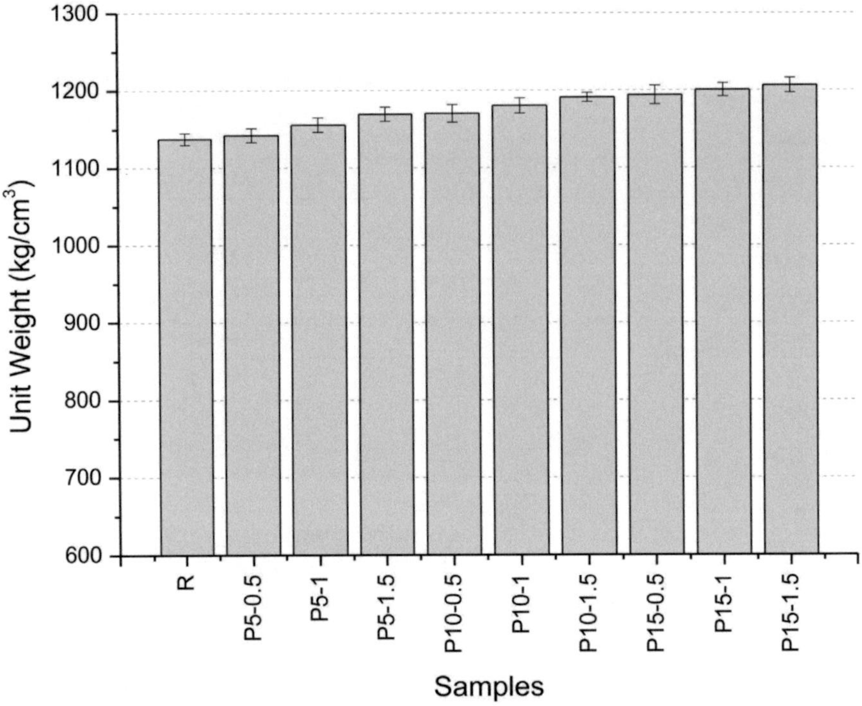


**Figure 3.**
*Unit weight of samples.*

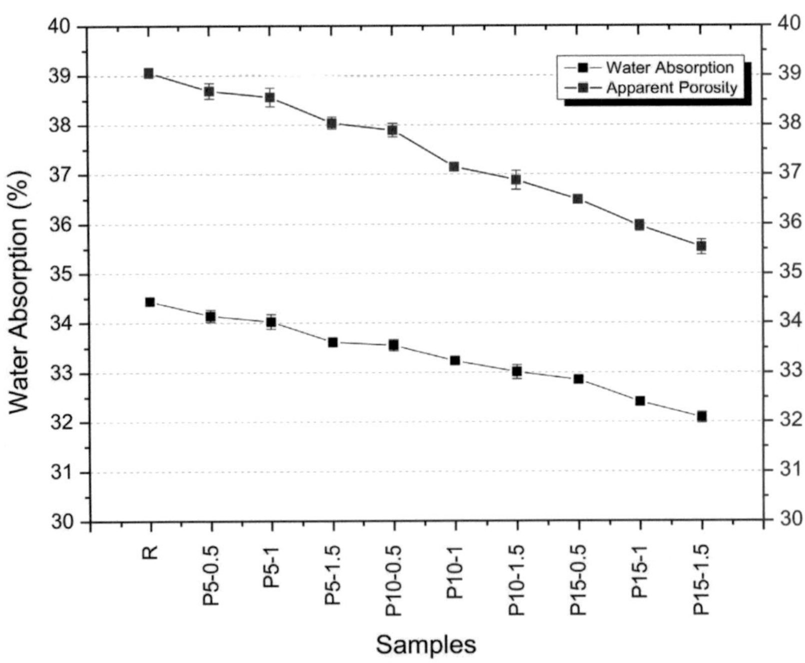


**Figure 4.**
*Water absorption and apparent porosity values.*

the highest unit weight value, the R sample gave the lowest. As the fiber amount and fiber length increase, there is a slight increase in the unit weights of the samples. This increase was at a level of 6% compared to the reference sample. Therefore, no significant change in unit weight values was observed using fiber. Although it is generally stated in the literature that studies on the use of PP fibers in concrete [29] and gypsum plaster [30] show a slight decrease in unit weight due to the density of the fibers being lower than the matrix, it is thought that the fibers used in this study have a reducing effect on the voids formed in the gypsum matrix. Thus, a slight decrease in unit weights was observed. The apparent porosity results also support these results.

**Figure 4** shows water absorption and apparent porosity values. Water absorption values of the samples changed between 32.09 and 34.44%. Porosity results were found to be in the range between 35.53 and 39.06%. The lowest porosity was obtained from the reference sample. Porosity was decreased with increase in fiber ratio. Same as porosity, water absorption values were also reduced with fiber ratio. These two results support each other. As stated before, the increase in fiber amount and size causes an increase in the unit weight of the samples. Since the density of tag pin fibers is lower than that of the gypsum matrix, this is attributed to a decrease in pore structure. Porosity results confirm this situation. There are other studies stating that there is a decrease in the pore structure with the use of PP fiber. For example, Ahmed et al. [31] reported in his study that the porosity of the sample produced only with cement was 21.2% after 48 h of hydration, while the porosity of the sample containing PP fiber was 19.2% after the same period. It is thought that this is because PP fibers have a high specific surface and thus accelerate the hydration kinetics. This way, the pores between the cement grains are quickly filled, and a denser matrix is obtained. In another study [32], PP fibers filled the pores in the mixture and reduced porosity, thanks to their flexibility and fineness. Although the fibers used in these studies are PP fibers with smaller sizes compared to the tag pin fibers used in this study, according to the study of Abousnina et al. [33], a study in which macro PP fibers were used, macro PP fibers also fill the gaps and cracks in the internal structure and thus finer pore structure has been reported. Since water absorption values are also related to the pore structure, it shows a similar trend.

**Figure 5** shows the ultrasonic pulse velocity (UPV) values of samples results. UPV values were between 2444 and 2601 m/s. UPV results are known to be strongly related to the material's pore structure. With the increase of pores or cracks in the matrix, the UPV decreases [34]. Herein, the lowest result was obtained from the reference sample. It is seen that UPV values increase as the fiber ratio and fiber length increase. This indicates that as the fiber ratio increases, the sizes of pores decrease. As discussed before, using tag pin fibers increased the unit weights of the samples and reduced their porosity values. As seen here, tag pin fiber has formed a relatively denser microstructure. According to these results, what is expected from the UPV test is that it shows an opposite trend with porosity results. As expected, the highest UPV value was obtained from the sample using 1.5% and 15 mm long tag pin fiber. This sample gave a 6.4% higher UPV value than the reference sample. In the study where Romero Gómez et al. [30] turned waste cloths into PP fiber, it was pointed out that UPV values decreased as the fiber ratio increased. The values ranged approximately between 1850 and 2250 m/s. In another study, the same author [35] evaluated waste fishing nets as fiber in gypsum-based composites and, this time, observed that UPV values increased as the amount of fiber increased. In a study on lightweight concretes containing PP fibers, Balgourinejad et al. [36] reported that using PP fibers did not significantly affect UPV results. A similar conclusion can be drawn considering the maximum increase in value observed in this study.

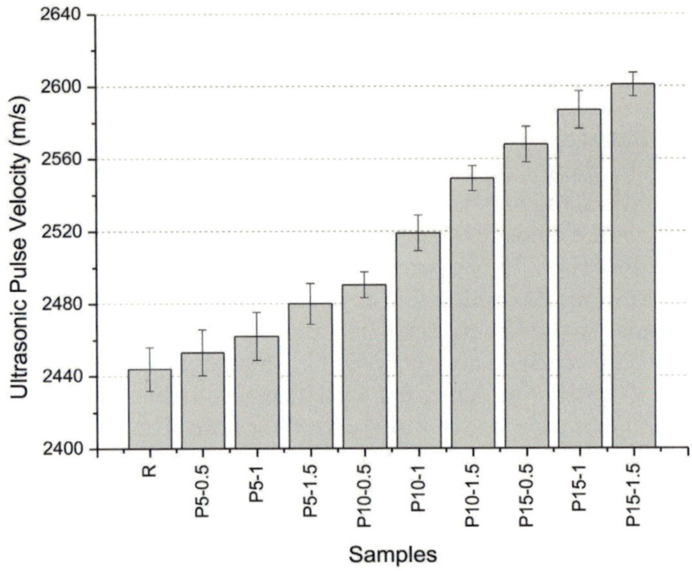


**Figure 5.**
*Ultrasonic pulse velocity values of samples results.*

**Figure 6** shows the water absorption by capillary suction after 10, 20, 30, 60, and 1440 min. The reference sample absorbed 102.6 g at 24 h. At the end of 24 h, the highest capillary water absorption value was obtained from the reference sample. In samples containing tag pin fiber, the highest value was obtained from the P5-0.5 sample (102.1 g), while the lowest value was obtained from the P15-1.5 sample

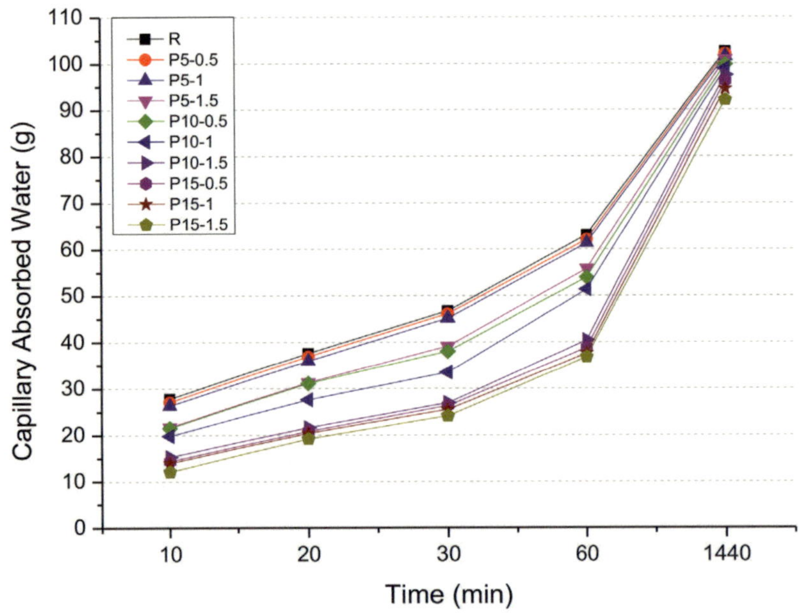


**Figure 6.**
*Results of water absorption by capillary suction after 10, 20, 30, 60, and 1440 min.*

(92.1 g). The increase in tag pin fiber usage rate and the use of longer fibers reduced capillary water absorption. While an approximately 10.2% reduction was achieved at the end of the 24-htest compared to the sample in which no fiber was used, switching from 5 mm fiber to 15 mm fiber at the same rate resulted in an average reduction of 7.1%. On the other hand, switching from 0.5% fiber to 1.5% fiber for the same length caused an average decrease of 2.58%. It is observed that the increase in fiber ratio is a more effective parameter than the increase in fiber length for improving capillary water absorption values. The capillary water absorption test results also support the previous porosity and UPV results. Although it is stated in the literature that the amount of open pores in mixtures containing PP fiber may be high [37], short-cut fibers will increase the capillarity effect in the gypsum plaster matrix [38]. However, it is also known that the fibers limit crack development and therefore may reduce water absorption capacity [37].

**Figure 7** shows the thermal conductivity coefficients of samples. Thermal conductivity coefficients varied between 0.388 and 0.413 W/mK. The lowest value was obtained from the reference sample. The highest value was obtained from the P15-1.5 sample. Thermal conductivity coefficients decreased as the tag pin fiber amount and fiber length increased. The reason for this decrease can be related to the pore structure of the samples. The general expectation for thermal conductivity coefficients is to show an opposite trend with the porosity results. The fact that using fibers reduces the number of pores also increases the thermal conductivity coefficient. Here, the P15-1.5 sample has a 6.4% higher thermal conductivity coefficient than the reference sample. In the literature, since PP fiber generally causes a slight increase in porosity in gypsum-based products, a decrease in thermal conductivity coefficients has also been observed. For example, the study of Gencel [39] achieved a 9.1% decrease in the thermal conductivity coefficient by using 0.5% PP fiber, while using 1% PP fiber increased this rate to 10%. In another study [40], the use of 0.5% PP fiber provided an 8.8% reduction, while the use of 1% PP fiber provided a 16.7% reduction. According

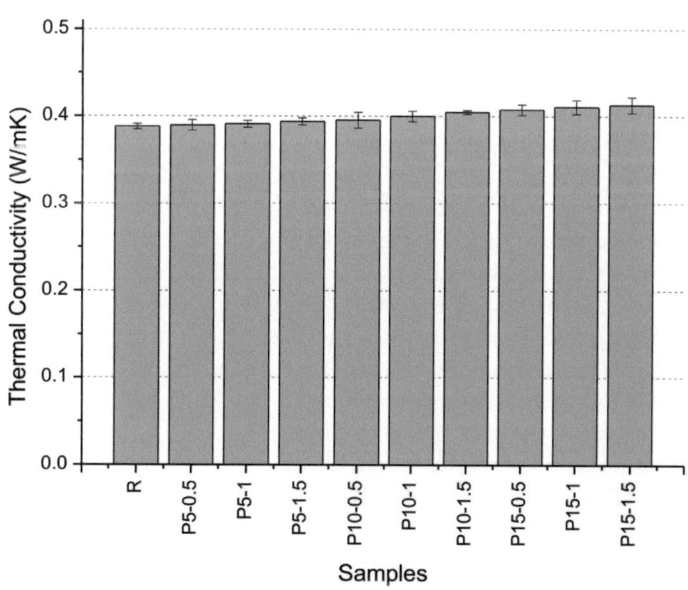


**Figure 7.**
*Thermal conductivity coefficients of samples.*

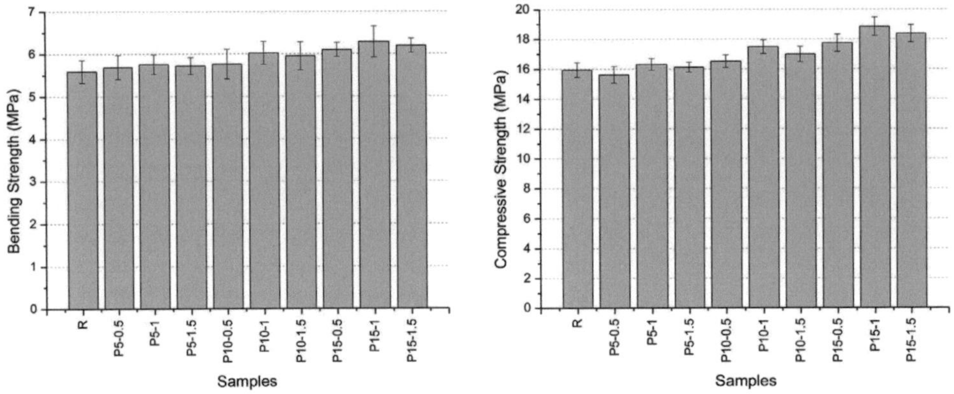


**Figure 8.**
*Results of bending and compressive strength tests.*

to these results, the use of tag pin fibers was found to be less effective compared to studies using PP fibers in the literature.

**Figure 8** shows the results of bending and compressive strength tests. The bending strengths of the samples altered between 5.59 and 6.28 MPa. The reference sample is the sample that gives the lowest bending strength. The use of tag pin fibers has increased the bending strength. In all three length groups, the lowest bending strength was obtained from the group containing 0.5% tag pin fiber, while the highest bending strength was obtained from the group containing 1.0% fiber. It can be stated that the optimum tag pin fiber usage rate among the mixtures applied within the scope of this study is 1%. When the sizes used were evaluated among themselves, the highest values were obtained when 15 mm length fiber was used. In this context, the highest bending strengths were obtained from the P15-1 sample. The value obtained here is 12.3% higher than the reference value. When 5 mm long tag pin fibers were used at 1%, an increase of only 2.3% was observed, but when 10 mm long fibers were used at 1%, an increase of 7.7% was observed. On the other hand, while using 5 mm fibers provides an average increase of 2.4%, these values are 5.7 and 10.7% for using 10 mm and 15 mm fibers, respectively. In a study in the literature involving the use of 0.5 and 1% PP fibers in gypsum-based composites [24], the bending strength increased by 6.25% with the use of 0.5% fiber and by 12.5% with the use of 1% fiber. In another study using the same ratios [40], using 0.5% PP fiber increased the bending strength by 1.7%, while using 1% fiber increased it by 21.4%. The fibers used in these examples are PP fibers specially produced for use in building materials such as concrete and mortar. Considering that the fibers used in this study are waste and are not PP fibers manufactured specifically to increase tensile strength, it can be said that the flexural strength increases obtained are at a satisfactory level.

Compressive strength values vary between 15.62 and 18.84 MPa. TS EN 13279-1 standard, which includes the properties of different gypsum plasters, states that the minimum compressive strength values of different gypsum products should be 2–6 MPa [41]. All samples produced within the scope of the study exceed these values. While the reference sample has a compressive strength of 15.94 MPa, the lowest strength value was obtained from the P5-0.5 sample. Except for this sample, all fiber-containing samples gave higher compressive strength than the reference sample. When samples containing fibers are evaluated among themselves, the trend is similar to that obtained from bending strength. Among the size groups, the highest compressive strength value was obtained from samples containing 1% tag pin fiber, while the lowest strength value was obtained from samples containing 0.5% fiber. At the same time, an increase in compressive strength was observed as the fiber length increased. The highest strength was obtained from the P15-1 sample. This value is 18.2% higher than that of the reference sample. The P5-0.5 sample, the only sample lower than that of the reference sample, gave a value only 2% lower. In their study, Gencel et al. [39] reported that the compressive strength of gypsum-based composites increased by 22% by using 0.5% PP fiber, and this rate increased to 55% with a 1% increase in PP fiber usage. In another study [24], it was reported that using 0.5% PP fiber provided a 3% increase in compressive strength, while using 1% PP fiber provided an increase of 6.7%. In a study where ladle slag and plaster were used together [42], it was observed that 2% PP fiber increased the compressive strength by approximately 36%. It is known that fiber increases compressive strength when the bond between fiber and matrix is good [42]. However, there are studies in the literature stating that compressive strength decreases with the use of PP fiber. For example, Durgun [40] observed that when 0.5% PP fiber was used in gypsum-based composites, there was a loss of compressive strength of 4.8%, and when 1% PP fiber was used, a loss of compressive strength of 9.7% occurred. This situation is attributed to increased porosity of the fibers [40] or poor interlocking between the fiber and the matrix [42]. This study observed that the use of fiber made the matrix structure denser and reduced porosity. It can be thought that the increases in compressive strength are related to this situation. As a result of both strength tests, it was seen that longer fiber length provided higher strengths. It is known that the length of the fiber affects the pull-out resistance [43]. This is attributed to the fact that as the embedded fiber length in the matrix increases, a more continuous surface is formed, and this increased surface provides more adhesion and friction, increasing the pull-out resistance and, therefore, the separation resistance of the matrix [44]. **Figure 9** shows various images from the fractured surfaces of various samples.

**Figure 10** shows the SEM image of a tag pin fiber on the fracture surface. When the fiber is examined in close view, it is seen that the surface of the fiber consists of two

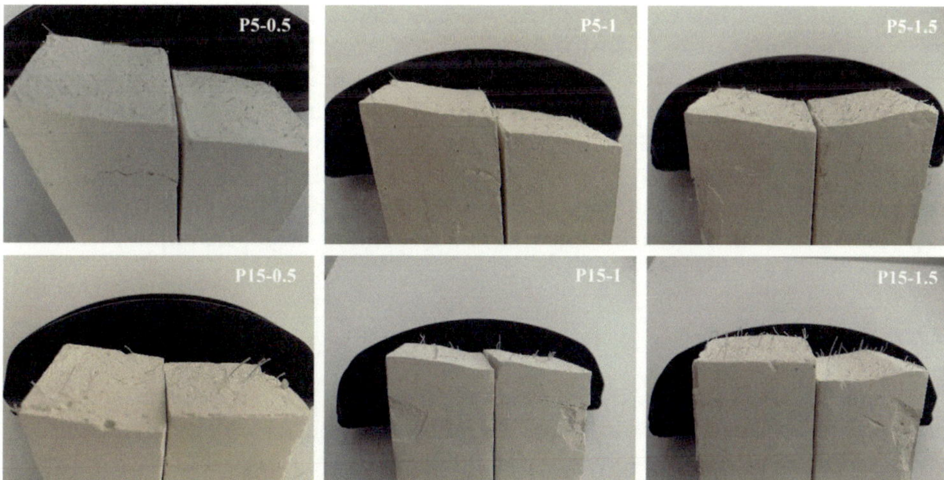


**Figure 9.**
*Various images from the fractured surfaces of various samples.*

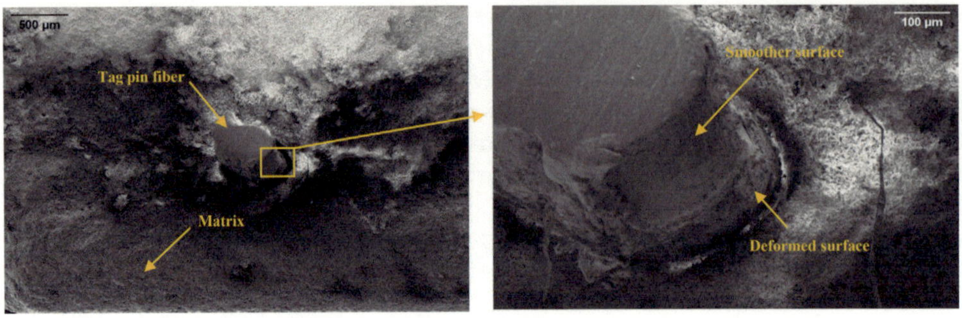


**Figure 10.**
*Scanning electron microscopy (SEM) image of a tag pin fiber on the fracture surface.*

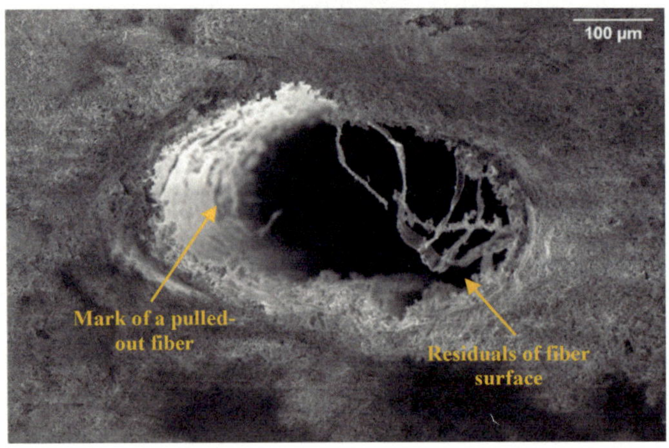


**Figure 11.**
*Scanning electron microscopy (SEM) image of a fiber mark in matrix.*

different textures. A section whose surface appears to be completely smooth and clean, as well as a deformed surface with matrix residues on it, can be seen. The presence of such matrix residues on the fiber surface is a sign of enough bonding [40, 42]. On the other hand, some deformations can occur on the fiber surface during pulling-out.

**Figure 11** shows an image of a fiber mark in the matrix as a hole left from a pulled-out fiber after the fracture. Here, at the edge of the pull-out mark, pieces that are thought to have been peeled off from the surface of the plastic fiber can be seen. An image like this suggests a good bonding between the tag pin fibers and the gypsum plaster matrix. The increases in compressive and bending strengths support this idea.

## 4. Conclusions

In this study, the use of waste tag pin fiber in gypsum plasters was investigated. According to the experimental studies, the following conclusions can be drawn:

The use of tag pin fibers increased the unit weight of the samples. It also reduced the apparent porosity values. This situation is also reflected in the UPV and water absorption results. It has been observed that increasing the amount and length of tag pin fibers has a slightly decreasing effect on porosity.

- Decreasing porosity increased the thermal conductivity coefficients. It was observed that the use of tag pin fibers slightly reduced the insulation properties of gypsum plaster samples.

- The use of tag pin fiber improved the mechanical properties of the samples. By increasing the amount and fiber length, an increase of up to 12.3% in bending strength and up to 18.2% in compressive strength was achieved. According to these results, it was observed that the optimum tag pin fiber usage rate was 1%, and the best results were obtained when 15 mm length fibers were used. The SEM images suggest that a sufficient bond between the tag pin fibers and the gypsum plaster matrix can be established.

## Author details

Ahmet Hayrullah Sevinç[1*], Muhammed Yasin Durgun[2] and Hayriye Hale Aygün[3]

1 Department of Construction, Elbistan Vocational School of Higher Education, Kahramanmaras Istiklal University, Kahramanmaras, Türkiye

2 Faculty of Architecture and Design, Department of Civil Engineering, Bartın University, Bartın, Türkiye

3 Department of Design, Vocational School of Technical Sciences, Kahramanmaras Sutcu Imam University, Kahramanmaras, Türkiye

*Address all correspondence to: ahmethayrullah.sevinc@istiklal.edu.tr

## References

[1] Dawood E, Mezal AM. The properties of fiber reinforced gypsum plaster. Journal of Scientific Research and Reports. 2014;**3**:1339-1347. DOI: 10.9734/jsrr/2014/7356

[2] Deng YH, Furuno T. Properties of gypsum particleboard reinforced with polypropylene fibers. Journal of Wood Science. 2001;**47**:445-450. DOI: 10.1007/BF00767896

[3] Arikan M, Sobolev K. The optimization of a gypsum-based composite material. Cement and Concrete Research. 2002;**32**:1725-1728. DOI: 10.1016/S0008-8846(02)00858-X

[4] Rostami R, Zarrebini M, Mandegari M, Mostofinejad D, Abtahi SM. A review on performance of polyester fibers in alkaline and cementitious composites environments. Construction and Building Materials. 2020;**241**:117998. DOI: 10.1016/j.conbuildmat.2020.117998

[5] Chung DDL, Zheng QJ. Electronic properties of carbon fiber reinforced gypsum plaster. Composites Science and Technology. 1989;**36**:1-6. DOI: 10.1016/0266-3538(89)90012-2

[6] Doshi S, Rathod J, Lad K, Parmar P, Patel P. Glass fiber reinforced gypsum panel use in building construction. Global Research and Development Journal for Engineering. 2018;**3**:47-51

[7] Hernandez Olivares F, Oteiza I, Villanueva LD. Experimental analysis of toughness and modulus of rupture increase of sisal short fiber reinforced hemihydrated gypsum. Composite Structures. 1992;**22**:123-137

[8] Li M, Pu Y, Thomas VM, Yoo CG, Ozcan S, Deng Y, et al. Recent advancements of plant-based natural fiber–reinforced composites and their applications. Composites. Part B, Engineering. 2020;**200**:1-52. DOI: 10.1016/j.compositesb.2020.108254

[9] Pacheco Torgal F, Jalali S. Cementitious building materials reinforced with vegetable fibres: A review. Construction and Building Materials. 2011;**25**:575-581. DOI: 10.1016/j.conbuildmat.2010.07.024

[10] Fantilli AP, Jóźwiak Niedźwiedzka D, Denis P. Bio-fibres as a reinforcement of gypsum composites. Materials (Basel). 2021;**14**(17):4830. DOI: 10.3390/ma14174830

[11] Jia R, Wang Q, Feng P. A comprehensive overview of fibre-reinforced gypsum-based composites (FRGCs) in the construction field. Composites. Part B, Engineering. 2021;**205**:108540. DOI: 10.1016/j.compositesb.2020.108540

[12] Sair S, Mandili B, Taqi M, El Bouari A. Development of a new eco-friendly composite material based on gypsum reinforced with a mixture of cork fibre and cardboard waste for building thermal insulation. Composites Communications. 2019;**16**:20-24. DOI: 10.1016/j.coco.2019.08.010

[13] Iucolano F, Liguori B, Aprea P, Caputo D. Evaluation of bio-degummed hemp fibers as reinforcement in gypsum plaster. Composites. Part B, Engineering. 2018;**138**:149-156. DOI: 10.1016/j.compositesb.2017.11.037

[14] Iucolano F, Boccarusso L, Langella A. Hemp as eco-friendly substitute of glass fibres for gypsum reinforcement: Impact and flexural behaviour. Composites.

Part B, Engineering. 2019;**175**:107073. DOI: 10.1016/j.compositesb.2019.107073

[15] Zargarnezhad H, Aghazadeh Mohandesi J. Experimental and multiscale numerical simulation of tensile strength of a randomly oriented short fiber composite in a brittle matrix. Computational Materials Science. 2013;**69**:533-541. DOI: 10.1016/j.commatsci.2012.08.033

[16] Suárez F, Felipe Sesé L, Díaz FA, Gálvez JC, Alberti MG. On the fracture behaviour of fibre-reinforced gypsum using micro and macro polymer fibres. Construction and Building Materials. 2020;**244**:118347. DOI: 10.1016/j.conbuildmat.2020.118347

[17] Hua S, Wang K, Yao X, Xu W, He Y. Effects of fibers on mechanical properties and freeze-thaw resistance of phosphogypsum-slag based cementitious materials. Construction and Building Materials. 2016;**121**:290-299. DOI: 10.1016/j.conbuildmat.2016.06.003

[18] Niinimäki K, Peters G, Dahlbo H, Perry P, Rissanen T, Gwilt A. The environmental price of fast fashion. Nature Reviews Earth & Environment. 2020;**1**:189-200. DOI: 10.1038/s43017-020-0039-9

[19] Shirvanimoghaddam K, Motamed B, Ramakrishna S, Naebe M. Death by waste: Fashion and textile circular economy case. Science of the Total Environment. 2020;**718**:137317. DOI: 10.1016/j.scitotenv.2020.137317

[20] Islam Rajib MM, Parvez MMH, Islam MS, Ahmed T, Islam MR. Complete garment costing with major cost breakdown. Journal of Textile Science & Fashion Technology. 2023;**9**(2):115-126. DOI: 10.4236/jtst.2023.92008

[21] Islam MA, Hossain MS, Badhon FF, Bhandari P. Performance evaluation of recycled-plastic-pin-supported embankment over soft soil. Journal of Geotechnical and Geoenvironmental Engineering. 2021;**147**:1-10. DOI: 10.1061/(asce)gt.1943-5606.0002528

[22] Khan MS, Kibria G, Hossain MS, Hossain J, Lozano N. Performance evaluation of a slope reinforced with recycled plastic pin. In: Proceedings of Geo-Congress; 3-7 March 2013; San Diego. California: Geo Congress; 2013. pp. 1733-1742. DOI: 10.1061/9780784412787.174

[23] Loehr JE, Bowders JJ, Owen JW, Sommers L, Liew W. Slope stabilization with recycled plastic pins. Transportation Research Record: Journal of the Transportation Research Board. 2000;**1714**:1-8. DOI: 10.3141/1714-01

[24] Gencel O, Del Coz Diaz JJ, Sutcu M, Koksal F, Alvarez Rabanal FP, Martinez Barrera G, et al. Properties of gypsum composites containing vermiculite and polypropylene fibers: Numerical and experimental results. Energy and Buildings. 2014;**70**:135-144. DOI: 10.1016/j.enbuild.2013.11.047

[25] Mohandesi JA, Sangghaleh A, Nazari A, Pourjavad N. Analytical modeling of strength in randomly oriented PP and PPTA short fiber reinforced gypsum composites. Computational Materials Science. 2011;**50**:1619-1624. DOI: 10.1016/j.commatsci.2010.12.020

[26] Jorillo PA, Shimizu G, Suzuki T. New technology approach to age-old waste material (natural fibers) for composites. In: Woolley GR, JJJM G, Wainwright PJ, editors. Waste Management Series. 1st ed. Amsterdam: Elsevier; 2000. pp. 45-56. DOI: 10.1016/S0713-2743(00)80017-3

[27] Zhu C, Zhang J, Peng J, Cao W, Liu J. Physical and mechanical properties of gypsum-based composites reinforced with PVA and PP fibers. Construction and Building Materials. 2018;**163**:695-705. DOI: 10.1016/j.conbuildmat.2017.12.168

[28] Wang Q, Cui Y, Xue J. Study on the improvement of the waterproof and mechanical properties of hemihydrate phosphogypsum-based foam insulation materials. Construction and Building Materials. 2020;**230**:117014. DOI: 10.1016/j.conbuildmat.2019.117014

[29] Karahan O, Atiş CD. The durability properties of polypropylene fiber reinforced fly ash concrete. Materials and Design. 2011;**32**:1044-1049. DOI: 10.1016/j.matdes.2010.07.011

[30] Romero Gómez MI, Pedreño Rojas MA, Pérez Gálvez F, Rubio de Hita P. Characterization of gypsum composites with polypropylene fibers from non-degradable wet wipes. Journal of Building Engineering. 2021;**34**:101874. DOI: 10.1016/j.jobe.2020.101874

[31] Ahmed A, Mahmoud AA, Elkatatny S. The effect of polypropylene fiber on the curing time of class G oil well cement and its mechanical, petrophysical, and elastic properties. Journal of Petroleum Exploration and Production Technologies. 2023;**13**:1181-1196. DOI: 10.1007/s13202-022-01601-2

[32] Liu Z, Yuan X, Zhao Y, Chew JW, Wang H. Concrete waste-derived aggregate for concrete manufacture. Journal of Cleaner Production. 2022;**338**:130637. DOI: 10.1016/j.jclepro.2022.130637

[33] Abousnina R, Premasiri S, Anise V, Lokuge W, Vimonsatit V, Ferdous W, et al. Mechanical properties of macro polypropylene fibre-reinforced concrete. Polymers (Basel). 2021;**13**:1-25. DOI: 10.3390/polym13234112

[34] Benaicha M, Jalbaud O, Hafidi Alaoui A, Burtschell Y. Correlation between the mechanical behavior and the ultrasonic velocity of fiber-reinforced concrete. Construction and Building Materials. 2015;**101**:702-709. DOI: 10.1016/j.conbuildmat.2015.10.047

[35] Romero Gómez MI, Silva RV, Flores Colen I, Rubio de Hita P. Mechanical performance of waste fishing net fibre-reinforced gypsum composites. Construction and Building Materials. 2023;**387**:131675. DOI: 10.1016/j.conbuildmat.2023.131675

[36] Balgourinejad N, Haghighifar M, Madandoust R, Charkhtab S. Experimental study on mechanical properties, microstructural of lightweight concrete incorporating polypropylene fibers and metakaolin at high temperatures. Journal of Materials Research and Technology. 2022;**18**:5238-5256. DOI: 10.1016/j.jmrt.2022.04.005

[37] İsafça Kaya T, Karakuzu K, Özen S, Doğangün A, Aghabaglou AM. Effect of polypropylene fiber and shrinkage reducing admixture utilization on water absorption of Khorasan mortar. Materials Today: Proceedings. 2022;**57**:730-733. DOI: 10.1016/j.matpr.2022.02.204

[38] Li Z, Wang X, Yan W, Ding L, Liu J, Wu Z, et al. Physical and mechanical properties of gypsum-based composites reinforced with basalt, glass, and PVA fibers. Journal of Building Engineering. 2023;**64**:105640. DOI: 10.1016/j.jobe.2022.105640

[39] Gencel O, Del Coz Diaz JJ, Sutcu M, Koksal F, Álvarez Rabanal FP, Martínez-Barrera G. A novel lightweight gypsum composite with diatomite and

polypropylene fibers. Construction and Building Materials. 2016;**113**:732-740. DOI: 10.1016/j.conbuildmat.2016.03.125

[40] Durgun MY. Effect of wetting-drying cycles on gypsum plasters containing ground basaltic pumice and polypropylene fibers. Journal of Building Engineering. 2020;**32**:101801. DOI: 10.1016/j.jobe.2020.101801

[41] Turkish Standardization Institute: TS EN 13279-1: Gypsum Binders and Gypsum Plasters - Part 1: Definition and requirements. 2009

[42] Nguyen H, Kinnunen P, Carvelli V, Mastali M, Illikainen M. Strain hardening polypropylene fiber reinforced composite from hydrated ladle slag and gypsum. Composites. Part B, Engineering. 2019;**158**:328-338. DOI: 10.1016/j.compositesb.2018.09.056

[43] Yoo DY, Park JJ, Kim SW. Fiber pullout behavior of HPFRCC: Effects of matrix strength and fiber type. Composite Structures. 2017;**174**:263-276. DOI: 10.1016/j.compstruct.2017.04.064

[44] Huang H, Gao X, Khayat KH, Su A. Influence of fiber alignment and length on flexural properties of UHPC. Construction and Building Materials. 2021;**290**:122863. DOI: 10.1016/j.conbuildmat.2021.122863

Chapter 6

# Low-Temperature Resistant Toughened Epoxy Resin Composite for Liquid Oxygen Tanks

*Baosheng Xu, Runze Jin, Ni Liu and Hui Wang*

## Abstract

Integrated molding of composite liquid oxygen tanks has become a crucial technology for aerospace carriers to shorten manufacturing cycles, improve carrying efficiency, and reduce production costs. Operating in an environment of −183°C, carbon fiber-reinforced resin composites face challenges such as extreme low-temperature service conditions and incompatibility with liquid oxygen due to the coupling effects of mechanical, thermal, and chemical fields. This chapter focuses on developing a modified epoxy resin matrix that is both ultralow temperature resistant and liquid oxygen compatible, essential for manufacturing composite liquid oxygen tanks. By improving the durability and fire resistance of the epoxy resin, this research introduces a novel dual-system macromolecular network interpenetration and interchain chemical crosslinking mechanism, improving mechanical properties at both room and ultralow temperatures. Moreover, two newly developed phosphorus/nitrogen-based reactive flame retardants are synthesized and added to the resin, notably improving both flame retardancy and compatibility with liquid oxygen. The resulting modified epoxy resin systems demonstrate superior mechanical properties at both room and ultra-low temperatures, making them suitable for manufacturing carbon fiber-reinforced composite materials for liquid oxygen tanks. The findings highlight the potential of these materials to meet the stringent requirements of aerospace applications.

**Keywords:** epoxy resin, composites, reinforcement and toughness, cryogenic mechanicals, flame retardant, liquid oxygen compatibility

## 1. Introduction

With the rapid evolution of composite material technology and its increasing integration into aerospace applications, the pursuit of lightweight propellant tanks through the utilization of linerless all-composite materials has emerged as a significant focus for space launch vehicle development. Among the array of materials considered for the fabrication of next-generation lightweight cryogenic propellant tanks, carbon fiber-reinforced resin-based composites (CFRP) have attracted significant interest as an exceptionally promising structural material. This interest stems from their exceptional specific strength and modulus, low density, thermal conductivity as well as outstanding corrosion resistance and fatigue properties [1]. However, the deployment of composite propellant tanks entails enduring severe operational conditions, including exposure to extreme temperatures, such as liquid hydrogen (−253°C) and liquid oxygen (−183°C) [2]. Moreover, there exists the critical challenge of mitigating the risk of material explosion attributed to oxygen incompatibility. Traditional bisphenol A-type epoxy resins exhibit deficiencies, such as inadequate mechanical properties and susceptibility to oxygen incompatibility at ultralow temperatures, rendering them unsuitable as resin matrixes for carbon fiber-reinforced resin-based composites. Hence, there is an urgent need to enhance the ultralow temperature mechanical properties and oxygen compatibility of epoxy resins.

This paper addresses several key challenges associated with epoxy resins, including high crosslinking density, brittleness, poor impact resistance, and molecular chain freezing at ultralow temperatures post-curing. To tackle these issues, a novel toughening and reinforcement modification approach is proposed. This approach involves the introduction of a dual-system macromolecular network interpenetration and interchain chemical crosslinking mechanism, aimed at enhancing the mechanical properties of epoxy resins both at room temperature and ultralow temperatures. Additionally, the paper presents the synthesis of two structurally distinct phosphorus/nitrogen reactive flame retardants through molecular structure design. The incorporation of these flame retardants improves the flame retardancy and liquid oxygen compatibility of the resins, resulting in a modified epoxy resin system suitable for use with liquid oxygen. The study further investigates the mechanical properties of the modified resins at both room and ultralow temperatures after the addition of flame retardants [3]. Furthermore, the paper explores the fabrication of carbon fiber-reinforced resin-based composites using both pure epoxy resin and modified epoxy resin as matrix materials. W-7011F plain weave fabric is employed as the reinforcing phase. The evolution of mechanical properties at room and ultralow temperatures is examined, and the influence of the resin matrix and ultralow temperature environment on the mechanical properties of the composite materials is analyzed through macroscopic and microscopic morphology analysis of composite fracture surfaces.

## 2. Preparation of experimental samples

### 2.1 Preparation of flame-retardant compounds

#### *2.1.1 Preparation of BSEA*

To prepare the desired compound, dissolve 5.00 g (18.65 mmol) of NEA in 30 mL of ethanol and transfer the solution into a 100 mL round-bottom flask. Subsequently, weigh 8.06 g (37.30 mmol) of DOPO and add it to the same flask. Stir the mixture and allow it to react at room temperature for 12 h. Following the reaction, gradually introduce the solution dropwise into deionized water to precipitate a yellow powder. Filter the suspension and transfer the resulting filter cake to a vacuum oven. Dry the material at 70°C to obtain a light yellow powder, BSEA, achieving a yield of 96.82%.

#### *2.1.2 Preparation of PBAH*

In a round-bottomed flask containing 50 mL of DMF, add 2.46 g (6 mmol) of BAPP, 1.47 g (12 mmol) of HB, and 2.59 g (12 mmol) of DOPO. Stir the reaction mixture using magnetic stirring at room temperature for 12 h. Subsequently, slowly introduce the reacted solution dropwise into deionized water to precipitate a yellow powder. Following filtration, transfer the filter cake to a vacuum oven and dry at 70°C to yield a light yellow powder, denoted as PBAH, with a high yield of 96.42%.

## 2.2 Preparation of modified epoxy resin and epoxy resin composite materials

### *2.2.1 Preparation of BCI-modified epoxy resin*

Bisphenol A epoxy resin and BCI were blended in an oil bath maintained at 80°C, following the predetermined ratio. Subsequently, the calculated amount of curing agent DDM was added, and mechanical stirring at 80°C was continued for approximately 30 minutes until complete dissolution of the DDM curing agent particles, resulting in a homogeneous and transparent brownish-red prepolymer. The obtained brownish-red transparent solution underwent vacuum defoaming treatment using a vacuum pump and defoaming bucket until all bubbles were eliminated. The defoamed prepolymer solution was then poured into a preheated stainless steel mold previously coated with a release agent. Subsequently, the mold was horizontally placed in a temperature-controlled oven and subjected to a gradient curing process at 100°C, 120°C, 140°C, 160°C, 180°C, and 200°C for 1 h. Upon completion of the curing process, the mold was cooled to room temperature within the oven to yield a series of EP/BCI composite materials. The resulting samples were polished using 200-mesh and 1000-mesh metallographic sandpaper, cleaned with ethanol, and dried, rendering them suitable for subsequent testing.

### *2.2.2 Preparation of BCI/BSEA-modified epoxy resin*

Add self-made BSEA with calculated amount to bisphenol A-type epoxy resin and stir mechanically at 150°C for 5 h to ensure sufficient reaction between BSEA and epoxy resin. Then cool the mixture to 80°C and add BCI in proportion. After BCI and the above mixture are evenly mixed, add a certain stoichiometric ratio of DDM. Then continue stirring at 80°C for about 30 minutes until DDM is completely dissolved, resulting in a uniform and transparent brownish-red prepolymer. Perform vacuum defoaming treatment on the obtained brownish-red transparent solution using a vacuum pump and defoaming bucket. After the bubbles are completely removed, pour the prepolymer solution into a preheated stainless steel mold pre-sprayed with release agent. Finally, place the mold in an oven and cure according to the following process: 100°C/1 h + 120°C/1 h + 140°C/1 h + 160°C/1 h + 180°C/1 h + 200°C/1 h. After curing, cool to room temperature with the oven to obtain EP/BCI/BSEA composite material.

### *2.2.3 Preparation of BCI/PBAH-modified epoxy resin*

Add self-made PBAH with calculated amount to bisphenol A epoxy resin and mechanically stir for 5 h at 150°C to ensure sufficient reaction between PBAH and epoxy resin. Then cool the mixture to 80°C and add BCI in proportion. After BCI and the above mixture are evenly mixed, add a certain stoichiometric ratio of DDM and continue stirring at 80°C until DDM is completely dissolved, obtaining a uniform and transparent brownish-red prepolymer. Perform vacuum defoaming treatment on the obtained brownish-red transparent solution using a vacuum pump and defoaming bucket. After the bubbles are completely removed, pour the prepolymer solution into a stainless steel mold pre-sprayed with a release agent. Finally, the mold is placed in an oven and cured according to the following process: 100°C/1 h + 120°C/1 h + 140°C/1 h + 160°C/1 h + 180°C/1 h + 200°C/1 h. After curing, it is cooled to room temperature in the oven to obtain EP/BCI/PBAH composite material.

#### *2.2.4 Preparation of carbon fiber-reinforced resin-based composite materials*

Carbon fiber-reinforced resin matrix composites were prepared using the hand layup method and vacuum bag forming process using plain woven W-7011F carbon fiber cloth as the reinforcing phase. The preparation process of its laminated board is as follows: 1. Cut the W-7011F plain weave carbon fiber cloth into a square fabric of about 300 mm × 300 mm, cut the demolding cloth, isolation film, adhesive felt into a size of 400 mm × 400 mm, and cut the vacuum bag into a size of 500 mm × 500 mm for later use; 2. Lay plain woven carbon fiber fabric layer by layer on a glass plate sprayed with a release agent, and sequentially, lay release fabric, isolation film, and adhesive felt on top. Preheat to 80°C in an oven; 3. According to the preparation method of modified resin, prefabricate the matrix resin solution and defoaming it for later use; 4. Transfer the preheated glass plate and carbon fiber cloth from the oven to the preheated heating plate and use a brush to brush the prefabricated resin solution layer by layer in one direction onto the carbon fiber cloth. After brushing, lay release cloth, isolation film, and adhesive felt on the carbon fiber cloth in sequence; 5. Seal the vacuum bag membrane with high-temperature resistant sealant and arrange a vacuum nozzle and guide pipe at one corner of the vacuum bag for the discharge of resin liquid and bubbles; 6. Transfer the packaged vacuum bag along with the glass bottom plate to the programmed temperature-controlled oven and connect the guide tube to the vacuum pump through a vacuum resin collector with a vacuum gauge. Turn on the vacuum pump and check if the vacuum bag is leaking. If there is any leakage, adjust the airtightness inside the vacuum bag by adjusting the high-temperature sealing adhesive to ensure that the vacuum degree of the system inside the vacuum bag reaches the highest; 7. Heat up and cure according to the preset curing process: 100°C/1 h + 12 0°C/1 h + 140°C/1 h + 160°C/1 h + 180°C/1 h + 200°C/1 h. After curing, cool to room temperature with the oven, remove the vacuum bag, and obtain composite laminates; and 8. Cut the prepared carbon fiber resin-based composite laminates according to the size of the sample to be tested using a water jet. The cut sample should be polished and reinforced with reinforcement plates as needed for subsequent testing.

## 3. Results and discussion

### 3.1 Flame retardancy and liquid oxygen compatibility of BCI-modified epoxy resin

#### *3.1.1 Mechanical properties of BCI-modified epoxy resin*

**Figure 1** illustrates the comprehensive mechanical properties of pure epoxy resin and BCI-modified resin at both room temperature and an ultralow temperature of 77 K [4]. This study evaluated several mechanical parameters, including tensile strength, elongation at break, bending strength, fracture toughness, and impact strength for both pure epoxy resin and EP/BCI composite materials. The radar plot

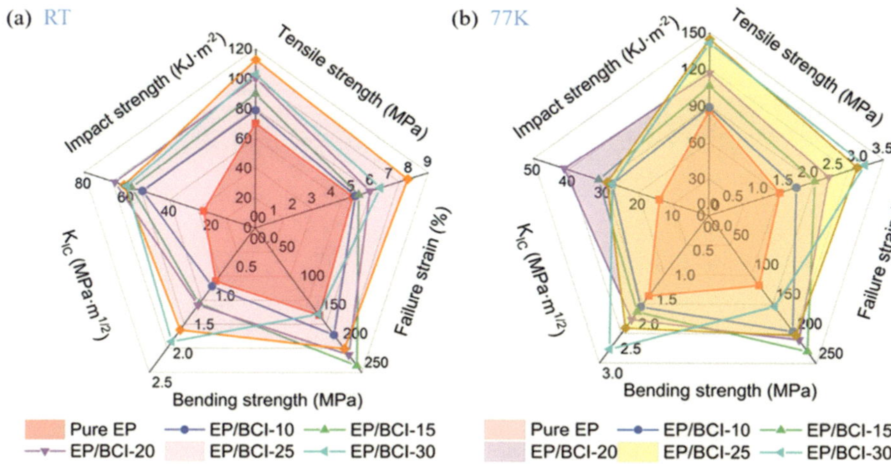


**Figure 1.**
*Comprehensive mechanical properties of EP/BCI composite material: (a) RT, (b) 77 K.*

in the figure quantitatively depicts these comprehensive mechanical properties. As demonstrated in **Figure 1(a)**, the EP/BCI composite materials exhibit significantly enhanced mechanical properties at room temperature compared to pure epoxy resin, with the EP/BCI-25 variant showing the most superior performance. Similarly, **Figure 1(b)** reveals that at an ultralow temperature of 77 K, the mechanical properties of the EP/BCI composite materials are markedly improved relative to those of the pure resin. When tensile strength and elongation at break are prioritized, EP/BCI-25 offers the best overall mechanical performance. Conversely, when impact strength is the main concern, EP/BCI-20 is found to be optimal. The toughening mechanism of BCI in epoxy resin involves two primary processes: Initially, the introduction of BCI modifies the crosslinking network within the epoxy resin. Concurrently, a series of chemical reactions occur during the curing of EP/BCI composites. Specifically, BCI integration fosters the formation of a dual macromolecular network system that interpenetrates with the epoxy matrix. This alteration increases the internal free volume within the resin system and decreases the crosslinking density, as detailed in **Figure 2**.

This enhancement increases the mobility of molecular chain segments, preventing their complete immobilization at ultralow temperatures. Consequently, some internal stresses within the resin system can still be alleviated through the dynamic movements of these segments. Additionally, during the curing process, BCI-modified epoxy resin undergoes a chemical reaction that facilitates interchain chemical crosslinking between BCI and the epoxy matrix. The incorporation of BCI into the epoxy resin system significantly augments the resin's resistance to crack propagation, thereby substantially improving its overall mechanical properties. Ultimately, the integration of BCI serves to toughen the epoxy resin and elevate its comprehensive mechanical performance.

### *3.1.2 Flame retardancy of BCI-modified epoxy resin*

The flame-retardant performance of EP/BCI composite materials was investigated. **Table 1** shows the flame-retardant rating and ultimate oxygen index of EP/BCI composite materials. From the table, it can be seen that the ultimate oxygen index of EP/BCI-25 reaches 29.6%, which is much higher than the ultimate oxygen index

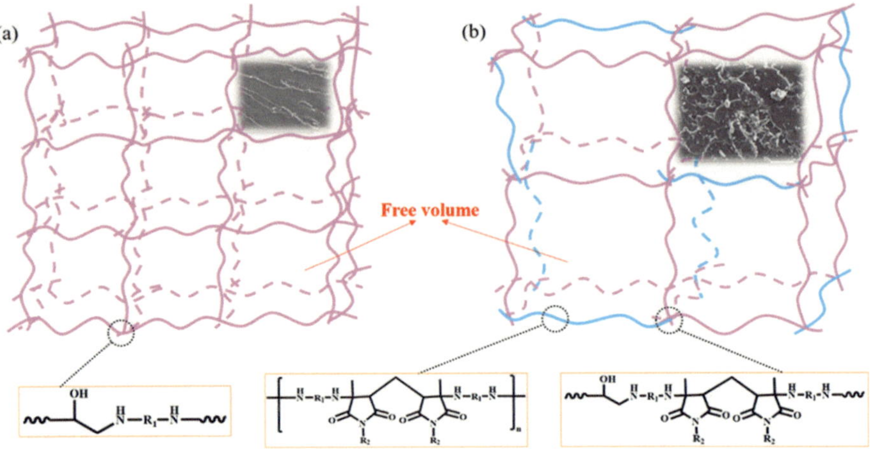


**Figure 2.**
*Schematic diagram of crosslinking network of materials: (a) pure epoxy resin, (b) EP/BCI composite material.*

| Samples | LOI (%) | UL-94 | HB | Dripping |
|---|---|---|---|---|
| Pure EP | 27.8 | NRa | HB-40 | Yes |
| EP/BCI-10 | 28.3 | NR | HB | No |
| EP/BCI-15 | 27.1 | NR | HB | No |
| EP/BCI-20 | 27.9 | NR | HB | No |
| EP/BCI-25 | 29.6 | NR | HB | No |
| EP/BCI-30 | 29.0 | NR | HB | No |

**Table 1.**
*Flame-retardant grade and ultimate oxygen index of EP/BCI composite materials.*

of pure epoxy resin (27.8%). Both pure epoxy resin and EP/BCI composite materials have not passed the vertical combustion test. The horizontal combustion level of pure epoxy resin is HB-40, while the horizontal combustion level of EP/BCI composite materials is HB. Moreover, there are no melted droplets during the combustion level determination of EP/BCI composite materials, indicating that the flame-retardant performance of BCI-modified resin is improved compared to pure epoxy resin but still cannot pass the vertical combustion test. Therefore, the flame-retardant performance of EP/BCI composite materials still needs to be improved [5].

### *3.1.3 Liquid oxygen compatibility of BCI-modified epoxy resin*

Through comprehensive analysis and research into the mechanical and flame-retardant properties of the EP/BCI-modified resin system under ambient and ultralow temperatures, it was discerned that the EP/BCI-25 variant exhibits superior mechanical and flame-retardant characteristics. Consequently, this section delves into the preliminary examination of liquid oxygen compatibility between pure epoxy resin and EP/BCI-25 *via* a 98 J liquid oxygen impact sensitivity assessment. **Table 2** presents the results of the liquid oxygen impact sensitivity test for both pure epoxy resin and EP/BCI-25. As depicted in the table, both pure epoxy resin and EP/BCI-25

| Samples | Experiment phenomena (number of times) | | | | Total number of tests | IRS (%) |
|---|---|---|---|---|---|---|
| | Burning | Explosion | Flash | Charring | | |
| Pure EP | 0 | 1 | 2 | 0 | 20 | 10.5 |
| EP/BCI-25 | 0 | 1 | 2 | 0 | 20 | 10.5 |

**Table 2.**
*Liquid oxygen shock sensitivity results of pure epoxy resin and EP/BCI-25.*

experienced one explosion and two flashes out of 20 liquid oxygen impact tests. Upon computation, the liquid oxygen impact sensitivity coefficients for both sample groups were determined to be 10.5%, signifying the incompatibility of both pure epoxy resin and EP/BCI-25 components with liquid oxygen. Consequently, in order to develop a composite resin matrix resilient to ultralow temperatures and compatible with liquid oxygen, it becomes imperative to flame-retardant modify the EP/BCI-25 composite material to ensure its compatibility with liquid oxygen.

## 3.2 Flame retardancy and liquid oxygen compatibility of BCI/BSEA-modified epoxy resin

### *3.2.1 Flame retardancy of BCI/BSEA-modified epoxy resin*

#### *3.2.1.1 The influence of BSEA on the flame-retardant grade and ultimate oxygen index of epoxy resin*

The influence of bisphenol S ethoxylate (BSEA) on the flame-retardant efficacy of epoxy resin was investigated using the limiting oxygen index (LOI) and UL-94 vertical combustion tests. The outcomes of these assessments are presented in **Table 3**. The LOI for pure epoxy resin and EP/BCI-25 were found to be 27.8% and 29.6%, respectively, and neither formulation passed the UL-94 rating test, indicating relatively poor flame-retardant properties. However, as the BSEA concentration increased, a noticeable improvement in LOI was observed. For example, EP/BCI/BSEA-2 (with 0.18 wt.% phosphorus) achieved an LOI of 30.7%, while EP/BCI/BSEA-3 (with 0.27 wt.% phosphorus) recorded an LOI of 31.8%, with both reaching a UL-94 classification of V-1. Furthermore, EP/BCI/BSEA-4 and EP/BCI/BSEA-5, containing 0.35 wt.% and 0.44 wt.% phosphorus respectively, attained LOIs of 32.1% and 32.8%, achieving the highest UL-94 rating of V-0. Notably, compared to pure

| Samples | LOI (%) | UL-94 | Dripping |
|---|---|---|---|
| Pure EP | 27.8 | NR | Yes |
| EP/BCI-25 | 29.6 | NR | No |
| EP/BCI/BSEA-2 | 30.7 | V-1 | No |
| EP/BCI/BSEA-3 | 31.8 | V-1 | No |
| EP/BCI/BSEA-4 | 32.1 | V-0 | No |
| EP/BCI/BSEA-5 | 32.8 | V-0 | No |

**Table 3.**
*LOI and UL-94 test results of EP/BCI/BSEA composite materials.*

epoxy resin, these BSEA-enhanced samples did not produce any melted droplets during combustion, demonstrating that the addition of BSEA significantly enhances the flame-retardant capabilities of epoxy resins.

#### *3.2.1.2 Analysis of flame-retardant mechanism of BSEA-modified epoxy resin*

The superior flame-retardant properties of BSEA-modified epoxy resin can be attributed to the synergistic effects of both gas-phase and condensed-phase mechanisms, as illustrated in **Figure 3**. Initially, during combustion, DOPO derivatives within BSEA release phosphorus-containing radicals such as $PO^{\bullet}$, $PO_2^{\bullet}$, and $HPO_2^{\bullet}$ in the gas phase. These radicals actively quench high-energy radicals, such as $H^{\bullet}$ and $OH^{\bullet}$, which are produced during the combustion of the resin. Furthermore, they can bind with the reactive sites of the resin matrix, effectively interrupting the radical chain reactions that propagate combustion. In addition to radical scavenging, BSEA also contributes to the generation of noncombustible gases such as $N_2$, $NO_2$, $NH_3$, $CO_2$, and $PH_3$ during thermal decomposition. These gases dilute the flammable gases emitted from the resin matrix, thereby impeding the exchange of oxygen and heat, which, in turn, diminishes the combustion intensity in the gas phase. Concurrently, in the condensed phase, phosphorus-rich residues promote the formation of a char layer, which acts as a physical barrier shielding the underlying resin from further combustion. This char layer expands and develops a porous structure due to the evolution of noncombustible gases, which disrupts the transfer of oxygen and heat, enhancing the material's flame retardancy. This dual-phase action renders the BSEA-modified epoxy resin highly effective at mitigating fire hazards.

### *3.2.2 Liquid oxygen compatibility of BCI/BSEA-modified epoxy resin*

#### *3.2.2.1 Sensitivity coefficient of liquid oxygen impact for BSEA-modified epoxy resin*

The liquid oxygen compatibility of epoxy resin modified with BSEA was assessed using the 98 J liquid oxygen impact sensitivity test, with an analysis of the epoxy

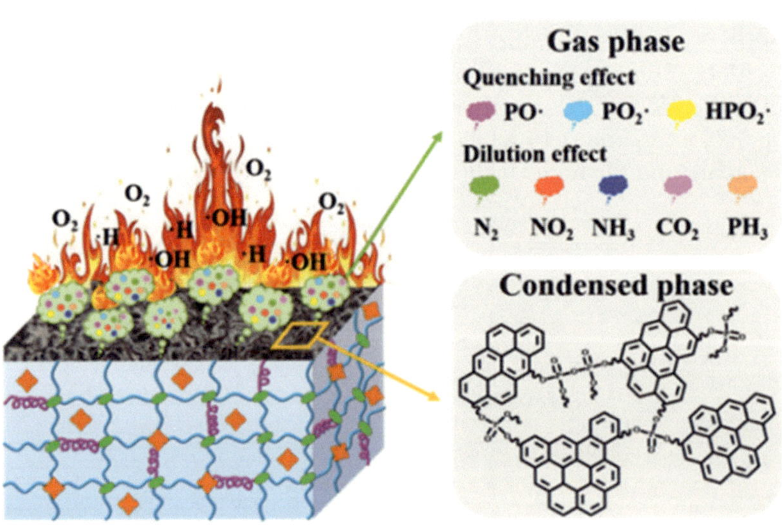


**Figure 3.**
*Flame-retardant mechanism diagram of BSEA-modified epoxy resin.*

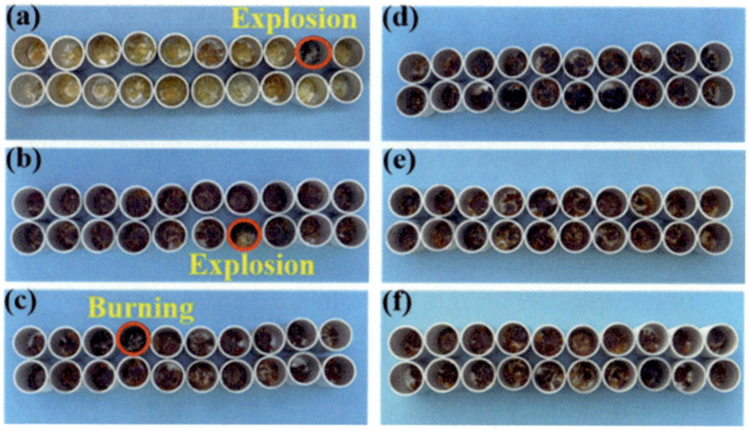


**Figure 4.**
*Sample images after liquid oxygen impact test: (a) pure epoxy resin, (b) EP/BCI-25, (c) EP/BCI/BSEA-2, (d) EP/BCI/BSEA-3, (e) EP/BCI/BSEA-4, and (f) EP/BCI/BSEA-5.*

resin's compatibility conducted *via* the impact reaction sensitivity coefficient. **Figure 4** presents digital images captured after the liquid oxygen impact testing, while **Table 4** provides details on the sensitivity reactions and liquid oxygen impact sensitivity coefficients of both pure epoxy resin and the modified resin throughout the liquid oxygen impact process. Observing **Table 4**, it becomes evident that both pure epoxy resin and EP/BCI-25 samples exhibited one explosion and two sparks during the liquid oxygen impact test. Upon calculation, the sensitivity coefficient for liquid oxygen impact was determined to be 10.5%. This suggests that both pure epoxy resin and EP/BCI-25 are incompatible with liquid oxygen, displaying a relatively high reaction strength with the oxidizer. During the liquid oxygen shock test, EP/BCI/BSEA-2 displayed one instance of combustion and one occurrence of coke formation, resulting in a sensitivity coefficient of 7.0%. Similarly, EP/BCI/BSEA-3 demonstrated two instances of coke formation during the liquid oxygen shock test, yielding a sensitivity coefficient of 4.0%. These results signify a gradual decrease in the sensitivity coefficient with the incorporation of BSEA. Furthermore, when the BSEA content reached 4 or 5 g, the modified resin exhibited no sensitivity reactions in 20 impact tests. Consequently, the liquid oxygen impact sensitivity coefficient was recorded as zero, indicating compatibility of the modified resin with liquid oxygen.

| Samples | Experiment phenomena (number of times) | | | | Total number of tests | IRS (%) |
|---|---|---|---|---|---|---|
| | Burning | Explosion | Flash | Charring | | |
| Pure EP | 0 | 1 | 2 | 0 | 20 | 10.5 |
| EP/BCI-25 | 0 | 1 | 2 | 0 | 20 | 10.5 |
| EP/BCI/BSEA-2 | 1 | 0 | 0 | 1 | 20 | 7.0 |
| EP/BCI/BSEA-3 | 0 | 0 | 0 | 2 | 20 | 4.0 |
| EP/BCI/BSEA-4 | 0 | 0 | 0 | 0 | 20 | 0 |
| EP/BCI/BSEA-5 | 0 | 0 | 0 | 0 | 20 | 0 |

**Table 4.**
*Liquid oxygen shock sensitivity results of EP/BCI/BSEA composite materials.*

These findings align with the flame-retardant performance of BSEA-modified resin, underscoring the efficacy of enhancing flame retardancy as a means to improve the liquid oxygen compatibility of epoxy resin.

##### *3.2.2.2 Analysis of liquid oxygen compatibility mechanism of BSEA-modified epoxy resin*

The mechanism underlying the compatibility of BSEA-modified epoxy resin with liquid oxygen is depicted in **Figure 5**. During the liquid oxygen impact test, the mechanical energy primarily transforms into internal energy. Under high-energy mechanical impact, the R−OH bond in the resin breaks down, resulting in the formation of $R^{\bullet}$ and $OH^{\bullet}$ radicals. Subsequently, $R^{\bullet}$ radicals react with liquid oxygen to generate $ROO^{\bullet}$ radicals, which can further react with $H^{\bullet}$ to form $COOH^{\bullet}$ radicals. These COOH radicals release $OH^{\bullet}$ radicals within the liquid oxygen environment, promoting the generation of C=O bonds. This formation of C=O bonds could be a crucial factor, influencing the epoxy resin's compatibility with liquid oxygen. Upon incorporating BSEA into the epoxy resin matrix, a significant amount of phosphorus-containing free radicals, such as $PO^{\bullet}$ and $HPO_2^{\bullet}$, are released during the liquid oxygen impact test.

These phosphorus-containing free radicals have the capability to capture high-energy free radicals, such as $H^{\bullet}$ and $OH^{\bullet}$, generated throughout the resin decomposition process. This capture interrupts the chain reaction of resin decomposition free radicals, akin to a flame-retardant mechanism. Consequently, this process ultimately renders the epoxy resin compatible with liquid oxygen.

Furthermore, the incorporation of BSEA enhances the comprehensive mechanical properties of the resin, bolstering its resistance to crack propagation. This enhancement results in greater dissipation of energy during liquid oxygen impact, thereby contributing to an improved extent of liquid oxygen compatibility for the resin.

### 3.3 Mechanical properties, flame retardancy, and liquid oxygen compatibility of BCI/PBAH-modified epoxy resin

#### *3.3.1 Flame retardancy of CI/PBAH-modified epoxy resin*

##### *3.3.1.1 The influence of PBAH on the flame-retardant grade and ultimate oxygen index of epoxy resin*

The influence of PBAH on the flame retardancy of epoxy resin was evaluated using the limiting oxygen index (LOI) and UL-94 vertical combustion tests, with the results detailed in **Table 5**. The LOI for the unmodified epoxy resin is recorded at 27.8%, which does not satisfy the criteria for the UL-94 rating, highlighting its relatively poor flame-retardant properties. However, with increasing PBAH concentrations, there is a noticeable enhancement in the LOI values. Specifically, the LOI for the EP/BCI/PBAH-3 formulation, containing 0.18 wt.% phosphorus, reaches 29.0%; for EP/BCI/PBAH-4 with 0.24 wt.% phosphorus, it increases to 30.1%; for EP/BCI/PBAH-5 with 0.30 wt.% phosphorus, it rises to 31.2%; and for EP/BCI/PBAH-6 with 0.35 wt.% phosphorus, it peaks at 31.7%. Remarkably, all four formulations achieved a V-0 classification in the UL-94 test, demonstrating the substantial flame-retardant efficacy of PBAH. Notably, compared to the baseline epoxy resin, the PBAH-enriched samples showed no signs of melting or dripping during combustion, underscoring the significant improvement in flame retardancy afforded by the incorporation of PBAH into the epoxy resin matrix.

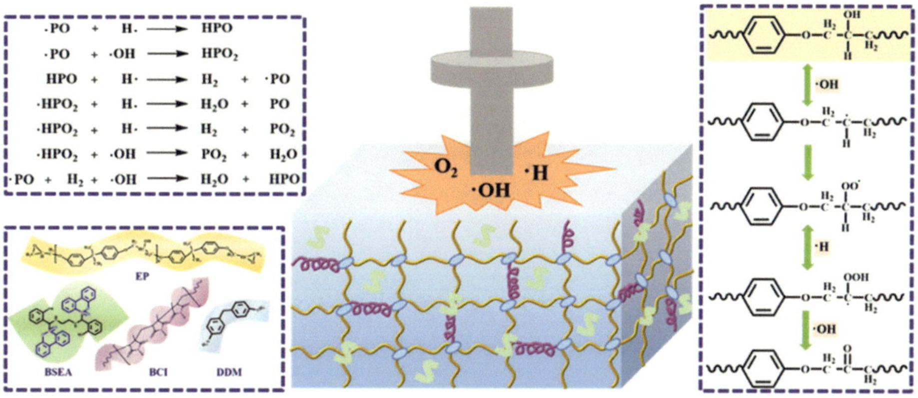


**Figure 5.**
*Mechanism diagram of liquid oxygen compatibility of BSEA-modified epoxy resin.*

| Samples | LOI (%) | UL-94 | Dripping |
|---|---|---|---|
| Pure EP | 27.8 | NR | Yes |
| EP/BCI/PBAH-3 | 29.0 | V-0 | No |
| EP/BCI/PBAH-4 | 30.1 | V-0 | No |
| EP/BCI/PBAH-5 | 31.2 | V-0 | No |
| EP/BCI/PBAH-6 | 31.7 | V-0 | No |

**Table 5.**
*LOI and UL-94 test results of EP/BCI/PBAH composite materials.*

### *3.3.1.2 Analysis of flame-retardant mechanism of PBAH-modified epoxy resin*

The superior flame-retardant performance of PBAH-modified epoxy resin can be attributed to the synergistic effects observed in both the gas-phase and condensed-phase mechanisms, as illustrated in **Figure 6**. In the gas phase, the incorporation of PBAH into the epoxy resin facilitates the release of phosphorus-containing radicals such as $PO^{\bullet}$, $PO_2^{\bullet}$, and $HPO_2^{\bullet}$ during combustion. These radicals effectively quench high-energy radicals such as $H^{\bullet}$ and $OH^{\bullet}$ produced during the resin combustion process. They also react with the active ends generated by the resin matrix, thereby disrupting the radical chain reaction that drives the combustion process. Moreover, PBAH generates noncombustible gases such as $N_2$, $NO_2$, $NH_3$, $CO_2$, and $PH_3$ during combustion. These gases serve as diluents, effectively isolating the combustion zone from ambient heat and oxygen, thereby reducing the intensity of combustion.

In the condensed phase, PBAH enhances the formation of a phosphorus-rich carbonaceous layer as the resin combusts. This layer swells due to the release of gases, forming a complex porous structure within the residual carbon. Such a layer acts as a physical barrier that impedes the transfer of heat and oxygen, thus further inhibiting the combustion of the underlying resin matrix. Additionally, the EP/BCI/PBAH formulations require a lower phosphorus content to achieve a V-0 rating, largely due to the phenyl ether structures in PBAH's molecular configuration. The incorporation of PBAH into the resin system increases the number of benzene rings, which elevates intra-chain rotational hindrance and enhances the thermal stability of the system.

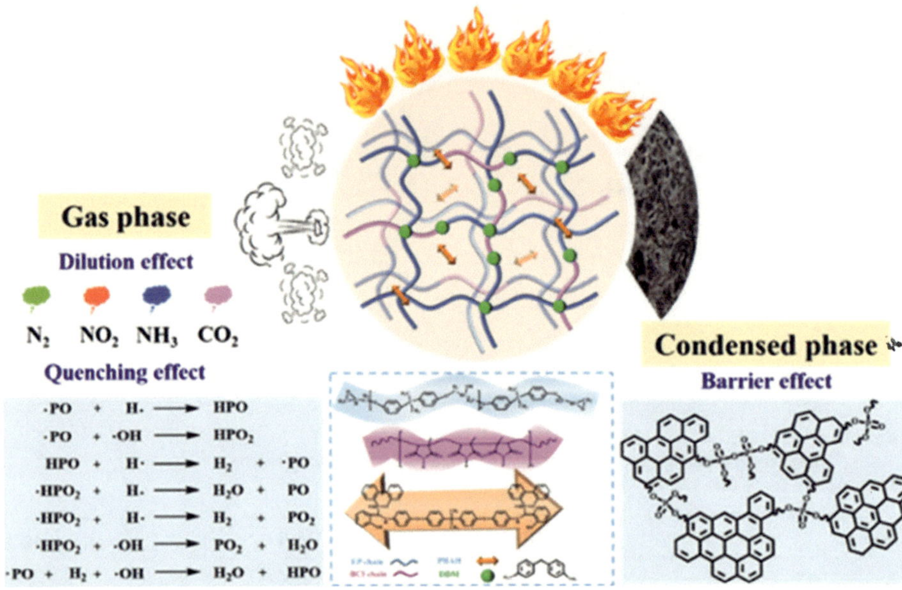


**Figure 6.**
*Flame-retardant mechanism diagram of PBAH-modified epoxy resin.*

Collectively, these results underscore that the combined gas-phase and condensed-phase actions of PBAH markedly elevate the flame retardancy of epoxy resin, surpassing that of BSEA in efficiency.

### *3.3.2 Liquid oxygen compatibility of CI/PBAH-modified epoxy resin*

#### *3.3.2.1 Sensitivity coefficient of liquid oxygen impact for PBAH-modified epoxy resin*

The liquid oxygen compatibility of PBAH-modified epoxy resin was evaluated using the 98 J liquid oxygen impact sensitivity test, and the analysis was conducted based on the impact reaction sensitivity coefficient. **Table 6** presents the sensitivity reactions and shock sensitivity coefficients of pure epoxy resin and PBAH-modified resin during the liquid oxygen shock process. As evident from **Table 6**, pure epoxy resin exhibited one explosion and two sparks during the liquid oxygen impact test, indicating its incompatibility and heightened reactivity

| Samples | Experiment phenomena (number of times) | | | | Total number of tests | IRS (%) |
|---|---|---|---|---|---|---|
| | Burning | Explosion | Flash | Charring | | |
| Pure EP | 0 | 1 | 2 | 0 | 20 | 10.5 |
| EP/BCI/PBAH-3 | 0 | 0 | 0 | 2 | 20 | 4.0 |
| EP/BCI/PBAH-4 | 0 | 0 | 0 | 1 | 20 | 2.0 |
| EP/BCI/PBAH-5 | 0 | 0 | 0 | 0 | 20 | 0 |
| EP/BCI/PBAH-6 | 0 | 0 | 0 | 0 | 20 | 0 |

**Table 6.**
*Liquid oxygen shock sensitivity results of EP/BCI/PBAH composite materials.*

with liquid oxygen. In contrast, EP/BCI/PBAH-3 demonstrated two instances of coking during liquid oxygen shock, with a calculated sensitivity coefficient of 4.0%. Similarly, EP/BCI/PBAH-4 displayed a coking phenomenon once during the liquid oxygen shock process, with a sensitivity coefficient of 2.0%, showcasing a gradual decrease in sensitivity coefficient with the incorporation of PBAH. Notably, at PBAH contents of 5 or 6 g, the modified resin showed no sensitivity reaction in 20 impact tests, with a liquid oxygen impact sensitivity coefficient of zero, indicating compatibility with liquid oxygen. These findings align with the flame-retardant performance of PBAH-modified resin, underscoring the efficacy of enhancing flame retardancy as a means to improve the liquid oxygen compatibility of epoxy resin.

#### *3.3.2.2 Analysis of liquid oxygen compatibility mechanism of PBAH-modified epoxy resin*

The sensitivity reaction diagram illustrating the liquid oxygen compatibility of epoxy resin is depicted in **Figure 7**, while the mechanism elucidating the liquid oxygen compatibility of PBAH-modified epoxy resin is delineated in **Figure 7(b)**. During the course of exposure to liquid oxygen, the majority of mechanical energy undergoes conversion into internal energy, engendering localized hotspots on the resin matrix surface. The resin surrounding these hotspots undergoes free radical chain reactions. Upon high-energy mechanical impact, the R−OH bond within the resin decomposes, yielding $R^•$ and $OH^•$ radicals. These $R^•$ radicals can interact with liquid oxygen, producing $ROO^•$ radicals, while concurrently reacting with $H^•$ to generate $^•COOH$ radicals. The release of $^•COOH$ radicals within the liquid oxygen environment, along with the formation of C=O moieties, may constitute pivotal factors influencing the epoxy resin's compatibility with liquid oxygen.

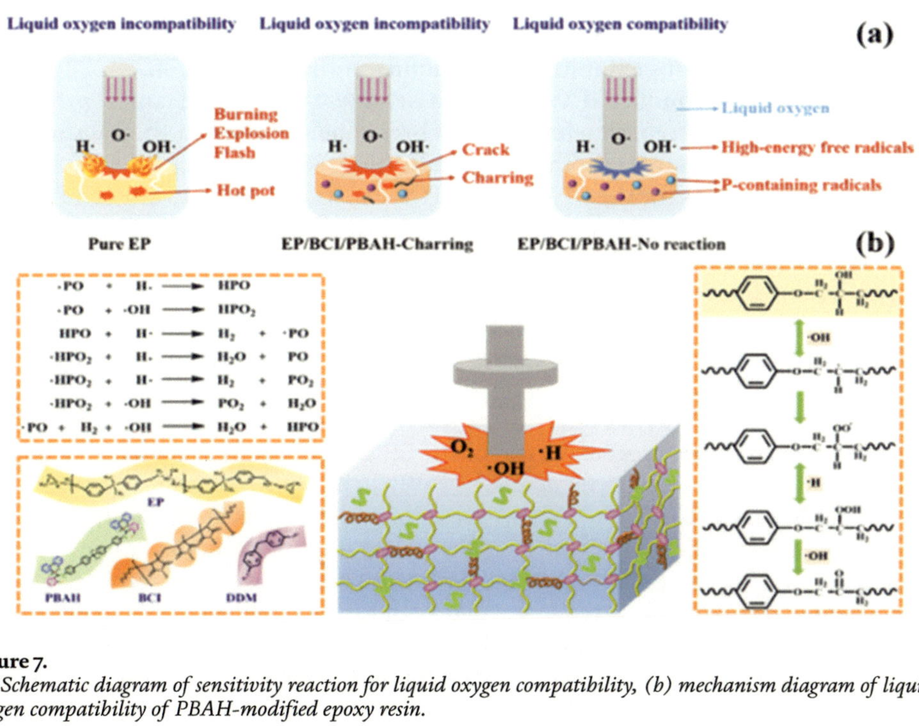


**Figure 7.**
*(a) Schematic diagram of sensitivity reaction for liquid oxygen compatibility, (b) mechanism diagram of liquid oxygen compatibility of PBAH-modified epoxy resin.*

Upon integration of PBAH into the epoxy resin matrix, a profusion of phosphorus-containing free radicals, such as PO and $HPO_2$, are liberated during the liquid oxygen impact evaluation. These phosphorus-containing radicals serve to sequester high-energy free radicals, such as H and $OH^•$, generated during the resin decomposition process. This interception disrupts the free radical chain reaction involved in resin decomposition, akin to the flame-retardant mechanism observed in BSEA. Ultimately, this intervention renders the epoxy resin compatible with liquid oxygen.

Furthermore, post-PBAH modification, the resin's overall mechanical properties are augmented, thereby bolstering its resistance to crack propagation. Moreover, a heightened dissipation of energy during exposure to liquid oxygen impacts is achieved. Collectively, these enhancements contribute to an improved liquid oxygen compatibility of the resin to a discernible degree.

## 3.4 Mechanical properties of carbon fiber-reinforced epoxy resin-based composite materials

### *3.4.1 Optimization of epoxy resin matrix*

In this study, we commenced by selecting an epoxy resin that showcased outstanding mechanical properties and compatibility with liquid oxygen from among those previously modified in the preceding chapters. This resin was chosen to serve as the matrix resin for fabricating carbon fiber resin-based composites. Among the candidates evaluated, both the pure epoxy resin and the EP/BCI-25 variant, which exhibited the most favorable comprehensive mechanical properties subsequent to BCI modification, were employed as control matrix resins.

Within the series of epoxy resin modifications involving BSEA, the EP/BCI/BSEA-4 and EP/BCI/BSEA-5 formulations demonstrated compatibility with liquid oxygen. Likewise, within the PBAH-modified series, the EP/BCI/PBAH-5 and EP/BCI/PBAH-6 formulations exhibited compatibility with liquid oxygen.

Consequently, we conducted a thorough assessment of the mechanical properties, including tensile strength, elongation at break, bending strength, fracture toughness, and impact strength, for both the EP/BCI/BSEA-4 and EP/BCI/BSEA-5 formulations, as well as for the EP/BCI/PBAH-5 and EP/BCI/PBAH-6 formulations. These evaluations were performed at both room temperature and ultralow temperatures. The outcomes of these assessments were visually depicted using radar plots, as depicted in **Figure 8**. Each radar plot delineates the area covered, serving as an indicator of the composite's comprehensive mechanical properties. These visual representations underscore the efficacy of the respective modifications in augmenting the overall performance of the materials under scrutiny.

Analysis of **Figure 8(a)** reveals that the radar area corresponding to EP/BCI/BSEA-4 at room temperature surpasses that of EP/BCI/BSEA-5, suggesting a superior comprehensive mechanical performance of EP/BCI/BSEA-4 under these conditions. Similarly, in **Figure 8(b)**, EP/BCI/BSEA-4 exhibits higher tensile strength, elongation at break, bending strength, fracture toughness, and impact strength at the ultralow temperature of 77 K compared to EP/BCI/BSEA-5. This is further corroborated by the substantially larger radar area associated with EP/BCI/BSEA-4, indicating its enhanced mechanical properties at ultralow temperatures in comparison to EP/BCI/BSEA-5. A thorough comparison demonstrates that EP/BCI/BSEA-4 outperforms EP/BCI/BSEA-5 in comprehensive mechanical properties, both at room temperature and

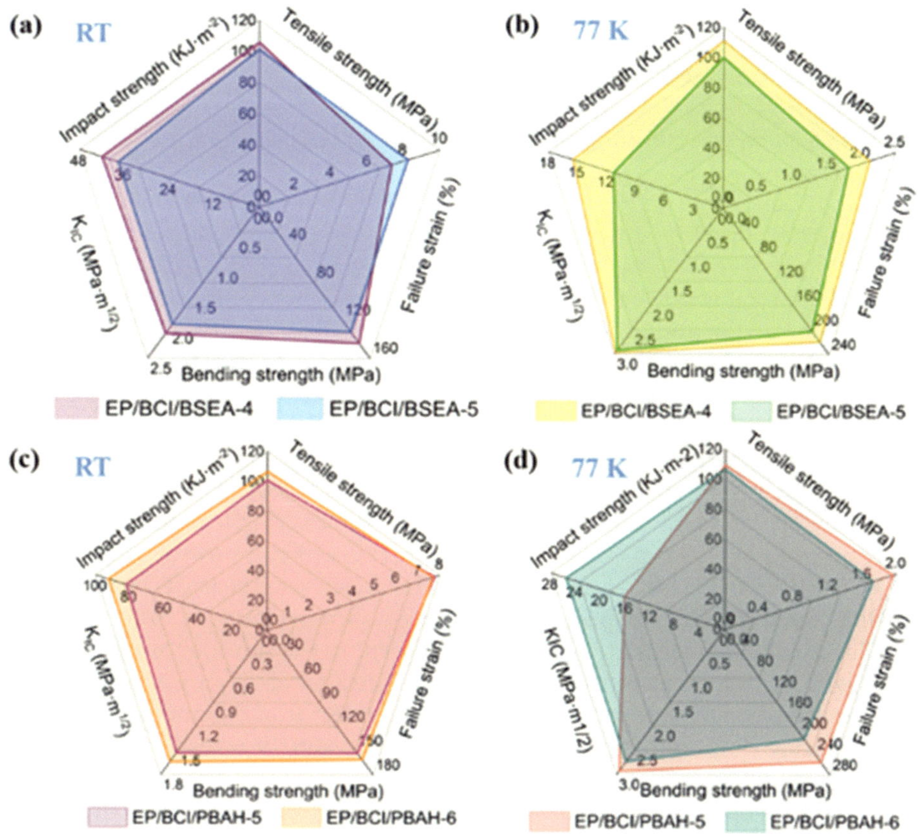


**Figure 8.**
*Comprehensive mechanical properties of EP/BCI/BSEA composite material: (a) RT, (b) 77 K; comprehensive mechanical properties of EP/BCI/PBAH composite materials: (c) RT, (d) 77 K.*

ultralow temperatures. Consequently, EP/BCI/BSEA-4 was selected as the matrix resin for BSEA-modified epoxy resin in the fabrication of carbon fiber resin-based composites.

In **Figure 8(c)**, it is evident that the radar cross-sectional area corresponding to EP/BCI/PBAH-6 at room temperature surpasses that of EP/BCI/PBAH-5, indicating the superior overall mechanical properties of EP/BCI/PBAH-6 under these conditions. Examining **Figure 8(d)**, it becomes apparent that EP/BCI/PBAH-5 exhibits superior tensile strength, elongation at break, bending strength, and fracture toughness compared to EP/BCI/PBAH-6. However, EP/BCI/PBAH-5's impact strength is marginally inferior to that of EP/BCI/PBAH-6. Consequently, in ultralow temperature environments, if emphasizing impact strength, EP/BCI/PBAH-6 demonstrates better overall mechanical properties than EP/BCI/PBAH-5. Conversely, if stringent impact strength requirements are not in place, precedence should be given to tensile strength, elongation at break, bending strength, and fracture toughness, where EP/BCI/PBAH-5 outperforms EP/BCI/PBAH-6. Following thorough comparison and deliberation, the final selection was narrowed down to four representative matrix resins: pure epoxy resin, EP/BCI-25, EP/BCI/BSEA-4, and EP/BCI/PBAH-5. The thermal expansion coefficients of these four representative matrix resins were characterized, and carbon fiber-reinforced resin matrix composites were fabricated using the hand layup method and vacuum bag forming process. Subsequently, the mechanical properties of the resulting carbon fiber-reinforced resin matrix composites were preliminarily explored and investigated.

### *3.4.2 Research on the mechanical properties of CFRP*

#### *3.4.2.1 Research on room temperature and ultralow temperature tensile properties of CFRP*

**Figure 9** illustrates the tensile properties of carbon fiber composites fabricated from pure epoxy resin and three modified resins—EP/BCI-25, EP/BCI/BSEA-4, and EP/BCI/PBAH-5—at room temperature. According to **Figure 9(a)**, the tensile strength of composites made from pure epoxy resin is recorded at 529.41 MPa. In contrast, composites fabricated with EP/BCI-25, EP/BCI/BSEA-4, and EP/BCI/PBAH-5 exhibit significantly enhanced tensile strengths of 733.87 MPa, 659.64 MPa, and 702.61 MPa, respectively, representing increases of 38.6%, 24.6%, and 32.7% over the pure epoxy resin. As depicted in **Figure 9(b)**, the elastic modulus of the pure epoxy resin composite stands at 36.86 GPa, while those of composites incorporating EP/BCI-25, EP/BCI/BSEA-4, and EP/BCI/PBAH-5 resins are measured at 47.06 GPa, 46.20 GPa, and 47.96 GPa, respectively, marking improvements of 27.7%, 25.3%, and 30.1%. These enhancements are primarily attributed to the superior tensile strength and elongation at the break of the modified resins compared to pure epoxy. Moreover, composites based on pure epoxy are susceptible to interfacial debonding between the resin and carbon fibers under tensile loads, which impairs the stress transfer capability of the resin matrix.

Concurrently, defects generated within the composite material led to the initial fracture of surrounding carbon fibers. As the tensile load intensifies, the load transitions across adjacent carbon fibers, culminating in an increasing number of fibers exhibiting fracture. This phenomenon prevents the fibers from exerting their strength simultaneously. However, in the modified EP/BCI-25, EP/BCI/BSEA-4, and EP/BCI/PBAH-5 carbon fiber composites, the notable enhancements in tensile strength and elongation at break of the resin matrix, as compared to those of the pure epoxy resin composites, bolster the matrix's ability to transmit stress. This enhancement enables a greater number of fibers to concurrently contribute to load-bearing,

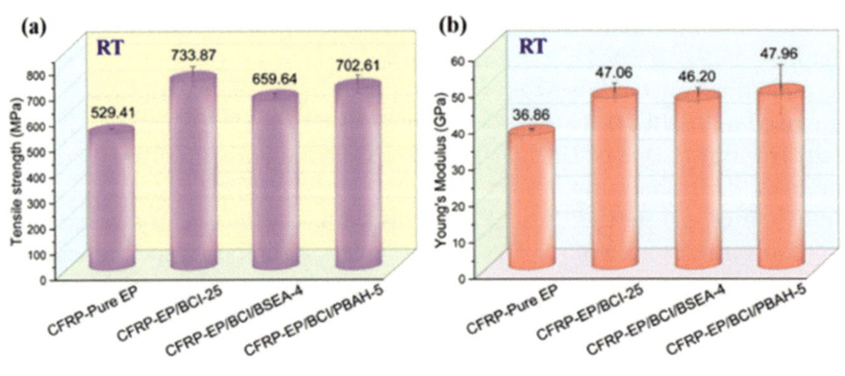


**Figure 9.**
*Tensile properties of carbon fiber resin-based composite materials at room temperature: (a) tensile strength, (b) elastic modulus.*

thereby significantly improving both the tensile strength and elastic modulus of the carbon fiber resin matrix composites. Additionally, the residual thermal stress, generated by temperature fluctuations during the transition from high temperature to room temperature at the conclusion of the curing process, adversely impacts the mechanical properties of the carbon fiber resin-based composite materials. The modified EP/BCI-25, EP/BCI/BSEA-4, and EP/BCI/PBAH-5 resins, featuring a reduced thermal expansion coefficient relative to pure epoxy resin, generate lower levels of residual thermal stress. This reduction in thermal stress contributes to an improvement in the mechanical properties of the modified carbon fiber composite materials.

**Figure 10** presents a comparison of the tensile properties of carbon fiber composites fabricated using pure epoxy resin and three modified resins: EP/BCI-25, EP/BCI/BSEA-4, and EP/BCI/PBAH-5 at ultralow temperatures. As illustrated in **Figure 10(a)**, the tensile strength of composites made with pure epoxy resin at ultralow temperatures registers at 493.13 MPa. In contrast, the tensile strengths of composites utilizing EP/BCI-25, EP/BCI/BSEA-4, and EP/BCI/PBAH-5 modified resins are 671.00 MPa, 637.11 MPa, and 666.63 MPa, respectively, representing increases of 36.1%, 29.2%, and 35.2% over the pure resin. **Figure 10(b)** shows that the elastic modulus of the pure epoxy resin carbon fiber composite is 36.19 GPa at ultralow temperatures, whereas the elastic moduli for composites modified with EP/BCI-25, EP/BCI/BSEA-4, and EP/BCI/PBAH-5 are 46.65 GPa, 45.06 GPa, and 43.62 GPa, respectively. These values indicate improvements of 28.9%, 24.5%, and 20.5% over the pure epoxy resin composite. These enhancements in tensile properties at ultralow temperatures are consistent with the improvements observed at room temperature, suggesting that the modified matrix resins effectively bolster the mechanical properties of the carbon fiber composites under these conditions. Comparing **Figures 9** and **10** reveals a slight reduction in the tensile strength and elastic modulus of carbon fiber composite materials when transitioning from room temperature to ultralow temperatures. Specifically, the tensile strength of pure epoxy resin and its modified counterparts—EP/BCI-25, EP/BCI/BSEA-4, and EP/BCI/PBAH-5—at ultralow temperatures decreased by 6.9%, 8.6%, 3.4%, and 5.1%, respectively, compared to room temperature. Similarly, the elastic modulus experienced reductions of 1.8%, 0.9%, 2.5%, and 9.0%, respectively under the same conditions. This decline can be primarily attributed to the incongruent thermal expansion coefficients between the carbon fiber and resin matrix in ultralow temperature environments. As temperatures decline from room temperature

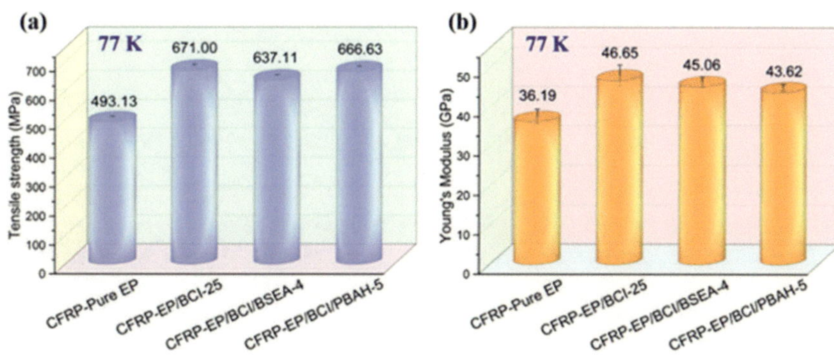


**Figure 10.**
*Tensile properties of carbon fiber resin-based composite materials at 77 K: (a) tensile strength, (b) elastic modulus.*

to −196°C, the resin matrix undergoes considerable shrinkage, inducing significant thermal stresses within the carbon fiber composite material. Furthermore, the disparity in deformation between the carbon fiber and resin matrix may provoke interface debonding, consequently leading to the formation of microcracks. Hence, the presence of thermal stress and microcracks within the composite materials constitutes the primary factors contributing to the observed decrease in tensile strength of carbon fiber-reinforced resin matrix composites in ultralow temperature environments. Comparing **Figures 9** and **10** reveals a slight reduction in the tensile strength and elastic modulus of carbon fiber composite materials when transitioning from room temperature to ultralow temperatures. Specifically, the tensile strength of pure epoxy resin and its modified counterparts—EP/BCI-25, EP/BCI/BSEA-4, and EP/BCI/PBAH-5—at ultralow temperatures decreased by 6.9%, 8.6%, 3.4%, and 5.1%, respectively, compared to room temperature. Similarly, the elastic modulus experienced reductions of 1.8%, 0.9%, 2.5%, and 9.0%, respectively under the same conditions. This decline can be primarily attributed to the incongruent thermal expansion coefficients between the carbon fiber and resin matrix in ultralow temperature environments. As temperatures decline from room temperature to −196°C, the resin matrix undergoes considerable shrinkage, inducing significant thermal stresses within the carbon fiber composite material. Furthermore, the disparity in deformation between the carbon fiber and resin matrix may provoke interface debonding, consequently leading to the formation of microcracks. Hence, the presence of thermal stress and microcracks within the composite materials constitutes the primary factors contributing to the observed decrease in tensile strength of carbon fiber-reinforced resin matrix composites in ultralow temperature environments.

#### *3.4.2.2 Research on the room temperature and ultralow temperature bending performance of CFRP*

**Figure 11** illustrates the bending performance of carbon fiber composite materials fabricated using both pure epoxy resin and modified variants—EP/BCI-25, EP/BCI/BSEA-4, and EP/BCI/PBAH-5—at ambient temperature. In **Figure 11(a)**, the bending strength of carbon fiber composites crafted with pure epoxy resin registers at 405.76 MPa, whereas those utilizing EP/BCI-25, EP/BCI/BSEA-4, and EP/BCI/PBAH-5 modified resins demonstrate bending strengths of 621.33 MPa, 647.72 MPa,

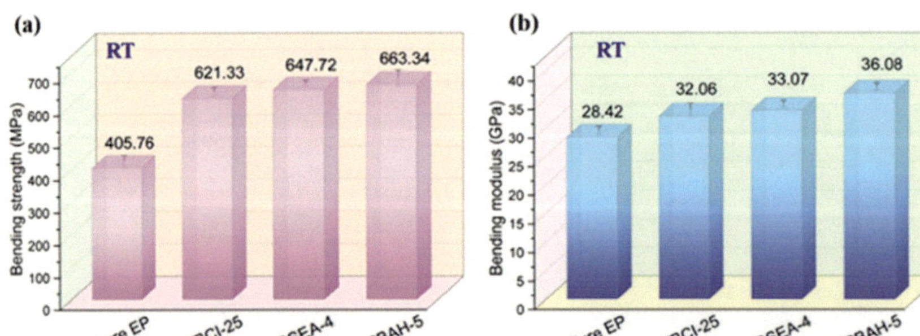


**Figure 11.**
*Bending performance of carbon fiber resin matrix composite at room temperature: (a) bending strength, (b) bending modulus.*

and 663.34 MPa, respectively. This denotes an enhancement in bending strength by 53.1%, 59.6%, and 63.5%, respectively, in comparison to pure epoxy resin-based carbon fiber composites. The predominant failure modes observed in carbon fiber resin-based composites during bending include tensile failure, compression failure, shear failure, and delamination. During three-point bending tests, the upper layer of the material experiences compressive stress, leading primarily to compression-induced failures, such as fracture and buckling, while the lower layer undergoes tensile stress, resulting mainly in fiber fracture, pullout, and delamination. Shear failure predominantly occurs in the intermediate layer of the material. Notably, during three-point bending tests, the onset of failure in all four types of carbon fiber resin-based composite materials primarily manifests on the tensile side.

When the stress state on the outermost side of the tensile zone of the composite material reaches the maximum load that the material can withstand, the composite material begins to fracture and fail. At the same time, the load transfers to the adjacent fiber layer, leading to further failure and failure of the material.

The tensile performance results indicate that the tensile strength of the modified resin composites EP/BCI-25, EP/BCI/BSEA-4, and EP/BCI/PBAH-5 at room temperature exceeds that of the pure epoxy resin composites. Additionally, the comprehensive mechanical properties of these modified resins enhance their stress transfer capabilities. Consequently, the bending strength of the modified composites EP/BCI-25, EP/BCI/BSEA-4, and EP/BCI/PBAH-5 at room temperature is significantly improved compared to that of the pure epoxy resin composites.

As demonstrated in **Figure 11(b)**, the bending modulus of the pure epoxy resin carbon fiber composite material at room temperature is 28.42 GPa. In contrast, the bending moduli of the carbon fiber composite materials modified with EP/BCI-25, EP/BCI/BSEA-4, and EP/BCI/PBAH-5 resins are 32.06 GPa, 33.07 GPa, and 36.08 GPa, respectively. Relative to the pure epoxy resin, these values represent increases of 12.8%, 16.7%, and 27.0%, respectively.

This enhancement in bending modulus is primarily attributed to two factors: First, the elastic modulus of the three modified resins within the carbon fiber composite materials is higher than that of the pure epoxy resin composite material on the tensile side. Second, on the compressive side, where the matrix resin predominantly bears the load, the elastic modulus of the modified resins also surpasses that of the pure epoxy resin. These two factors synergistically contribute to the higher bending modulus of the carbon fiber composite materials composed of EP/BCI-25, EP/BCI/BSEA-4, and EP/BCI/PBAH-5 modified resins at room temperature compared to the pure epoxy resin composite material.

**Figure 12** presents the load-displacement curves for carbon fiber composite materials prepared with pure epoxy resin and modified resins—EP/BCI-25, EP/BCI/BSEA-4, and EP/BCI/PBAH-5—during bending tests. **Figure 12(a)** and **(b)** indicate that at both room temperature and ultralow temperatures, the four types of carbon fiber composites display similar load-displacement curves. Initially, the load and displacement exhibit a linear relationship until reaching the failure load, after which the load decreases and displays a sawtooth-like wave pattern. This pattern is primarily due to the failure of the outermost fibers on the tensile side after the material reaches its ultimate load, leading to a reduction in load-bearing capacity. Subsequently, the load is transferred to the adjacent fiber layer through the resin, which still retains some capacity to resist external loads, resulting in serrated fluctuations in the load. Comparing **Figure 12(a)** and **(b)** reveals that these serrated fluctuations are more pronounced at ultralow temperatures than at room temperature, suggesting that

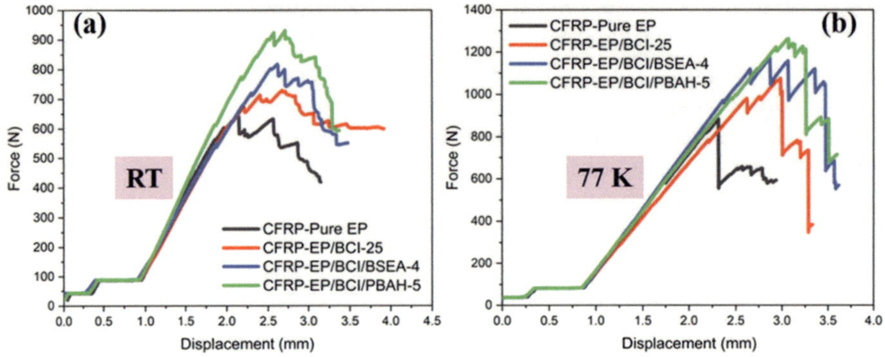


**Figure 12.**
*Bending load-deflection curve of carbon fiber resin-based composite material: (a) room temperature, (b) ultralow temperature.*

carbon fiber composites possess greater resistance to external loads in ultralow temperature environments. This increased resistance is attributed to resin matrix shrinkage in ultralow temperature environments, enhancing the material's resistance to interlayer shear failure. Therefore, in ultralow temperature conditions, carbon fiber composites primarily exhibit interlayer shear-induced delamination failures, which manifest as pronounced sawtooth fluctuations on the load-displacement curves.

#### *3.4.2.3 Research on room temperature and ultralow temperature interlaminar shear properties of CFRP*

As illustrated in **Figure 13**, the interlayer shear properties of carbon fiber composite materials fabricated using both pure epoxy resin and modified resins—EP/BCI-25, EP/BCI/BSEA-4, and EP/BCI/PBAH-5—are delineated. In **Figure 13(a)**, it is evident that the interlayer shear strength of carbon fiber composites crafted with pure epoxy resin at room temperature registers at 29.92 MPa. In contrast, the interlayer shear strength of carbon fiber composites incorporating EP/BCI-25, EP/BCI/BSEA-4, and EP/BCI/PBAH-5 modified resins reaches 37.66 MPa, 49.57 MPa, and 45.77 MPa, respectively, marking increases of 25.9%, 65.7%, and 53.0% over pure epoxy resin composites. This enhancement primarily stems from the substantially improved

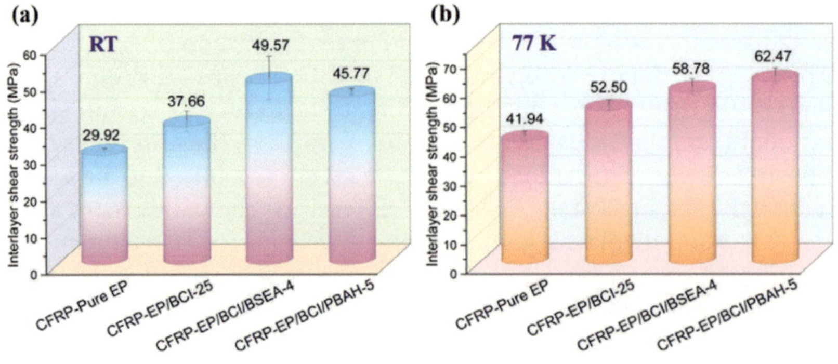


**Figure 13.**
*Interlayer shear strength of carbon fiber resin-based composite materials: (a) room temperature, (b) ultralow temperature.*

fracture elongation and toughness exhibited by the modified EP/BCI-25, EP/BCI/BSEA-4, and EP/BCI/PBAH-5 resins compared to pure epoxy resins. The fortified resin matrix facilitates more efficient load transfer, enhances crack propagation resistance, and bolsters delamination resistance in carbon fiber composites. Consequently, the interlayer shear strength of carbon fiber composites employing EP/BCI-25, EP/BCI/BSEA-4, and EP/BCI/PBAH-5 modified resins at room temperature surpasses that of pure epoxy resin composites. Notably, the incorporation of flame-retardants BSEA and PBAH in EP/BCI/BSEA-4 and EP/BCI/PBAH-5 carbon fiber composites has led to varying degrees of improvement in shear strength compared to EP/BCI-25. This observation underscores the efficacy of flame-retardants BSEA and PBAH in augmenting the interfacial bonding between the resin matrix and carbon fibers.

In **Figure 13(b)**, the interlayer shear strength of carbon fiber composites formulated with pure epoxy resin under ultralow temperature conditions is reported as 41.94 MPa. Conversely, carbon fiber composites incorporating EP/BCI-25, EP/BCI/BSEA-4, and EP/BCI/PBAH-5 modified resins exhibit interlayer shear strengths of 52.50 MPa, 58.78 MPa, and 62.47 MPa, respectively—reflecting increments of 25.2%, 40.2%, and 49.0% over pure epoxy resin composites. The interlayer shear failure mechanism observed in carbon fiber composite materials at ultralow temperatures mirrors that observed at room temperature. Under external loading, the resin matrix transfers stress loads to adjacent carbon fiber layers *via* shear stress across the interface layer, ultimately resulting in interlayer failure of the composite material. The fortified resin matrix facilitates enhanced load transmission, thereby effectively bolstering the interlayer shear strength of the composite material. A comparative analysis between **Figure 13(a)** and **(b)** reveals that the interlayer shear strength of the same carbon fiber composite material is notably elevated in ultralow temperature environments compared to room temperature environments. The interlayer shear strength of pure epoxy resin, EP/BCI-25, EP/BCI/BSEA-4, and EP/BCI/PBAH-5 modified resin composite materials at ultralow temperature is increased by 40.2%, 39.4%, 18.6%, and 36.5%, respectively, compared to room temperature environment. This is mainly attributed to the increased toughness of the modified resin due to the shrinkage of the resin molecular crosslinking network in the ultralow temperature environment, which enables the resin matrix to transmit more loads when subjected to external loads. Moreover, the interface bonding energy between the resin and fibers is enhanced due to resin shrinkage at ultralow temperatures, resulting in strong resistance to interlayer shear failure of composite materials. The reddest result is that the interlayer shear strength of pure epoxy resin, EP/BCI-25, EP/BCI/BSEA-4, and EP/BCI/PBAH-5 resin matrixes in the ultralow temperature environment is better than that in the room temperature environment.

## 4. Conclusion

The integration of BCI into epoxy resin significantly enhances its mechanical properties at room and cryogenic temperatures, attributed to a novel dual-network interpenetrating and interchain chemical crosslinking mechanism. This modification results in substantial improvements in tensile strength, elongation, flexural properties, fracture toughness, and impact resistance. Reactive phosphorus/nitrogen flame retardants, BSEA and PBAH, developed from specific chemical compounds, further augment the resin's flame retardancy and liquid oxygen compatibility, achieving a UL-94 V-0 rating and higher oxygen index. Optimally formulated carbon fiber composites exhibit superior mechanical performance over pure epoxy resin, particularly at cryogenic temperatures, due to improved interfacial bonding. This research addresses fundamental challenges in developing cryogenic and liquid oxygen compatible composite materials, offering vital theoretical and technical insights for advancing aerospace material technologies.

## Author details

Baosheng Xu[1,2*], Runze Jin[1,2], Ni Liu[1,2] and Hui Wang[1,2]

1 Institute of Advanced Structure Technology, Beijing Institute of Technology, Beijing, PR China

2 Beijing Key Laboratory of Lightweight Multi-Functional Composite Materials and Structures, Beijing Institute of Technology, Beijing, PR China

*Address all correspondence to: xubaosheng211@163.com

## References

[1] Li Y, Guan H, Bao Y, et al. Ni0.6Zn0.4Fe2O4/Ti3C2Tx nanocomposite modified epoxy resin coating for improved microwave absorption and impermeability on cement mortar. Construction and Building Materials. 2021;**310**:125213. DOI: 10.1016/j.conbuildmat.2021.125213

[2] Liu N, Wang H, Xu B, et al. Cross-linkable phosphorus/nitrogen-containing aromatic ethylenediamine endowing epoxy resin with excellent flame retardancy and mechanical properties. Composites Part A: Applied Science and Manufacturing. 2022:107145. DOI: 10.1016/j.compositesa.2022.107145

[3] Li JL, Wang C, Lu KY. Enhanced cryogenic mechanical properties and liquid oxygen compatibility of DOPO-containing epoxy resin reinforced by epoxy-grafted polysiloxane. Polymer Bulletin. 2020;**77**:3429-3442. DOI: 10.1007/s00289-019-02931-8

[4] Liu N, Wang H, Ma B, et al. Enhancing cryogenic mechanical properties of epoxy resins toughened by biscitraconimide resin. Composites Science and Technology. 2022;**220**:109252

[5] Liu N, Wang H, Wang S, et al. Reactive phosphaphenanthrene aromatic ether diamine endowing epoxy resin with excellent fire resistance, liquid oxygen compatibility and cryogenic mechanical properties. Reactive and Functional Polymers. 2023;**183**:105500. DOI: 10.1016/j.reactfunctpolym.2023.105500

Chapter 7

# Co-Extruded Wood-Plastic Composites: Their Structure, Properties, and Applications

*Kaimeng Xu, Huanbo Wang and Tian Liu*

## Abstract

Wood-plastic composites (WPCs) are innovative materials prepared by mixing wood fibers, thermoplastics, and additives through extrusion, injection molding, or compression molding processes. Co-extruded WPCs (Co-WPCs) are multilayer composites, in which regular WPCs are coated with a protective shell layer through coextrusion. The core-shell structure of Co-WPCws provides a way to overcome the shortcomings of WPCs by modifications specific to the composite surfaces. With the development of coextrusion technology, Co-WPCs show promise to become one of the important leading directions of the WPC industry. Based on the special characteristics of the core-shell structure, the properties of Co-WPCs were reviewed in this chapter, including mechanical properties, dimensional stability, weather resistance, flame retardancy, etc. Furthermore, the applications of Co-WPCs were comprehensively presented. Finally, the problems and challenges in the development of Co-WPCs were put forward, and the key points of future research were also expounded.

**Keywords:** wood-plastic composites, coextrusion, core-shell structure, properties, application

## 1. Introduction

Wood-plastic composites (WPCs) are composite materials made primarily from polar wood fibers and non-polar thermoplastic polymers, produced through melting, compounding, including extrusion, hot pressing, injection molding, and other molding processes [1–3]. WPCs have excellent comprehensive performance such as anti-corrosion, durability, waterproofing, and moisture resistance, as well as low cost, environmental friendliness, and excellent mechanical properties. They have been widely used in areas, such as architectural decoration, interior design, outdoor landscaping, and so on. The main drawbacks of WPCs are the moisture absorption caused by the hydrophilic wood flour, which increases the degradation and loss of properties of the WPCs [4, 5]. Thermoplastic materials are typically hydrophobic, hence WPCs with injection-molded plastic-rich surfaces exhibit less water absorption and improved weathering resistance [6, 7]. Therefore, applying a layer rich in plastic to WPCs is a viable way to lessen the composites' absorption of moisture and improve their resilience to weathering. WPCs' flammability because of the organic nature of its primary ingredients (thermoplastics) is another drawback. Adding fire retardants to bulk composites is the standard approach of reducing the flammability of WPCs [8]; however, adding a large amount of fire retardants frequently degrades the mechanical qualities of WPCs and raises costs. The inside condition of the composites has no bearing on the initial damage to WPCs since external factors, including moisture, ultraviolet (UV) radiation, and combustion sources, primarily affect the surface of WPCs. Hence, surface modification rather than bulk materials should be the main emphasis of WPC protection. Moreover, by incorporating additives only into the outer layer, the necessary amount of additives is reduced. As a result, applying a functional protective surface layer to WPCs is a cost-effective and efficient way to enhance their durability.

Coextrusion is a process where two or more polymeric materials are combined through a feedblock or die to create a single multilayer structure [9, 10]. Various polymers can be merged to form a unified structure with improved properties. Coextrusion has emerged as a leading plastic processing technology for producing customizable multilayer composites, commonly utilized in manufacturing tubing, multilayer sheets, wire coatings, films, and core-shell profiles [11]. Co-WPCs are typically manufactured with core-shell structures to introduce a cap layer to WPCs, as illustrated by the Co-WPC sample shown in **Figure 1**. The coextrusion system comprises two extruders for the core layer and shell layer melts, along with a specifically designed die for transforming the two melts into the desired core-shell profile. The Co-WPCs combine the different performance advantages of the surface layer and core layer materials. The optimization of overall performance within the core-shell structure enhances its added value and provides diverse functionalities. Therefore, Co-WPCs have the potential to replace traditional wood-plastic composites in various application fields, particularly in outdoor environments with extreme conditions and areas with special functional requirements.

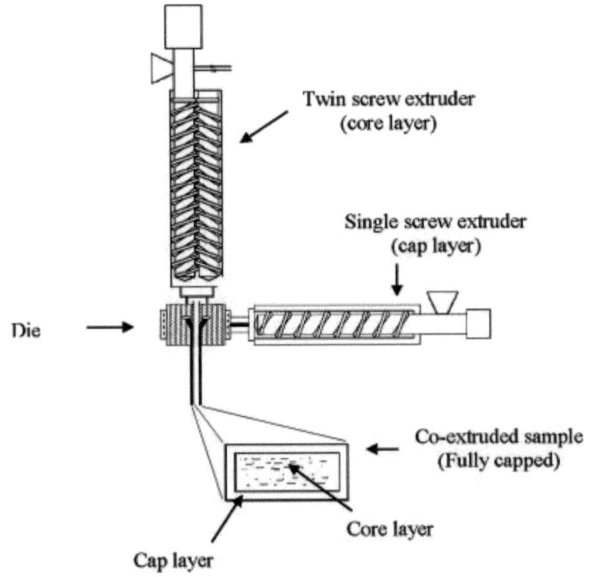


**Figure 1.**
*Scheme of the coextrusion system used to produce Co-WPCs [12].*

Researchers first prepared WPCs with a core-shell structure, known as Co-WPCs, using coextrusion technology in 2007 [13]. With the development of the wood-plastic industry over the years, Co-WPCs have matured and gradually become a research hotspot in the field of WPCs, showing promising development potential. In recent years, while demand for traditional WPC products has slowed down, the development of Co-WPC products has surged, with its market share gradually increasing to over 50%. Co-WPCs are likely to become the mainstream trend in the WPC industry. The presence of the shell layer not only imparts multifunctionality to Co-WPCs but also provides them with diverse decorative characteristics such as rich colors and surface textures. Therefore, Co-WPCs are expected to exhibit a trend toward diversification in development. This chapter summarizes the structure design, properties, and applications of Co-WPCs.

## 2. Structure design of Co-WPCs

As people's requirements for the performance of WPCs and their products continue to increase, traditional single-structure wood-plastic composites can no longer meet the special performance, functionality, and appearance requirements of wood-plastic products. Co-WPCs can combine materials with different characteristics to fully leverage the inherent properties of various materials, producing products for specific purposes with short molding cycles and low energy consumption. Different types of composite products can be produced by co-extruding materials with different structures. Coextrusion of materials with different characteristics can give a product excellent properties from several different materials. By employing suitable structure design, high added-value co-extruded wood-plastic composite products can be manufactured to meet various performance and application requirements.

### 2.1 Fully and non-fully wrapped Co-WPCs

The shell layer of Co-WPCs is typically fully coated on the surface of the core layer, blocking the contact between the core layer material and the external environment, and serving a protective function for the core layer. Research has shown that the mechanical and thermal properties of Co-WPCs with fully coated surfaces are superior to those of semi-coated materials [14]. Therefore, adopting a semi-coated structure on the surface to save costs is not worthwhile. The typical structures of fully and non-fully wrapped Co-WPCs are shown in **Figure 2**.

The conventional non-fully wrapped Co-WPCs often rely solely on the adhesive properties between the surface layer and the core layer after heating and melting, but the limitations of adhesion between inner and outer layer materials, as well as differences in material properties such as shrinkage/expansion rates and weather resistance in different usage environments, often lead to detachment and peeling between the core layer and surface layer at the transitional region of semi-coated bonding, resulting in issues such as shortened lifespan and product vulnerability. By incorporating a fixed groove design, the coating material extends beyond just attaching to the board surface, reaching into the interior cross-section of the core layer material, significantly increasing the contact area, and effectively preventing coating layer detachment. These embedded non-fully wrapped Co-WPCs, compared to traditional non-fully wrapped Co-WPCs, offer the advantage of structural stability over traditional fully coated boards and cost savings.

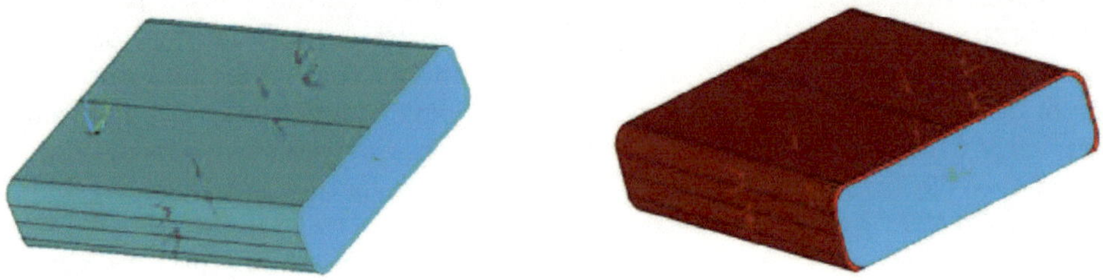

**Figure 2.**
*Fully (left sample) and non-fully (right sample) wrapped Co-WPCs [14].*

Wood-plastic boards produced by traditional extrusion methods exhibit low impact toughness, susceptibility to breakage, and other shortcomings. Moreover, as the content of wood flour increases, the strength and modulus initially rise to a certain extent before sharply declining, with a more pronounced decrease in toughness. Lowering the wood flour content can improve impact toughness to a certain degree, but the strength and modulus may not meet requirements. Therefore, the issue of balancing impact toughness and strength modulus in wood-plastic boards remains a challenge for the wood-plastic industry. In response to this problem, by producing sandwich-structured Co-WPCs with a low-density polyethylene core structure, compared to wood-plastic composites without extruded core structures, bending strength can be increased by 10–50%, bending modulus by 5–50%, and impact strength by 20–80%. This enhancement makes it possible to simultaneously strengthen and toughen wood-plastic composite materials mechanically.

### 2.2 Co-WPCs with solid wood core

Currently, Co-WPCs prepared through extrusion technology have relatively low mechanical strength. Solid wood materials exhibit higher comprehensive mechanical performance compared to wood-plastic composite materials; however, they are prone to warping and deformation after absorbing moisture, easily decay, and are challenging to utilize efficiently. In response to these issues, scholars have proposed using wood-plastic composite materials as the surface layer and fast-growing artificial forest wood and reconstituted wood (referred to as solid wood) as the core material. Co-WPCs with solid wood core formed through coextrusion exhibit better anti-creep properties, non-brittleness, lightweight, high strength, cost-effectiveness, and excellent durability. In comparison to traditional Co-WPCs, Co-WPCs with solid wood core-only require one extruder and can be prepared by a specific coextrusion die and solid wood transport device to achieve simple or complex cross-sectional Co-WPCs (**Figure 3**) [15]. Co-WPCs with solid wood core demonstrate excellent comprehensive performance and are environmentally friendly, with costs falling between wood-plastic and wood-based multilayer boards, providing high cost-effectiveness. By designing mortise and tenon structures, the outer-shell-layer polyvinyl chloride (PVC) wood-plastic composite (WPVC) and the inner-core-layer laminated veneer lumber (LVL) could form a mechanical interlock, which could effectively improve the interface bonding property between the two layers [16]. They are not only suitable as upgraded replacement products for traditional preservative wood and engineered panels but also exhibit outstanding advantages in high-quality building doors and windows, construction templates, large-span wooden components, architectural

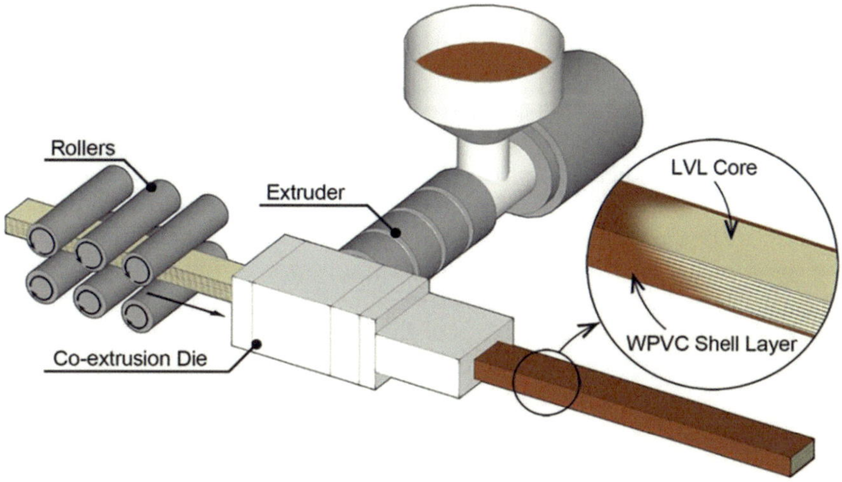


**Figure 3.**
*Schematic of the fabrication of Co-WPCs with solid wood core [15].*

components, multifunctional wall structures, green buildings, and other applications requiring materials with high environmental, load-bearing, waterproof, moisture-resistant, decay-resistant, and weather-resistant properties, offering broad application prospects.

### 2.3 Specially structured Co-WPCs

To enhance the performance of Co-WPCs to meet different application scenarios, other specially structured Co-WPCs have been developed. The specially structured Co-WPCs include Co-WPCs with multilayered core, Co-WPCs with compartmentalized surface structures, Co-WPCs with metal core, co-extruded wood-plastic-rebar composite materials, and co-extruded core foaming composite materials. However, these special structural co-extruded wood-plastic composite materials have been reported only in some patents and have yet to appear in academic papers and practical applications.

## 3. Main properties of Co-WPCs

Compared to ordinary WPCs, Co-WPCs exhibit better toughness, but there will be varying degrees of reduction in strength and modulus, especially when different functional fillers are added to the shell layer, which significantly impact the mechanical performance. Mechanical performance has always been a key issue in Co-WPC research. Additionally, the higher polymer content in the shell layer results in higher temperature dependency of Co-WPCs, making it necessary to study the influence of external environmental factors (such as temperature, humidity, and external forces) on their dimensional stability. Furthermore, during long-term indoor and outdoor use, Co-WPC products have certain requirements for flame resistance and weather resistance. It is also important to consider the balance between flame resistance, weather resistance, and other properties to avoid significant decreases in individual performance. In summary, this section systematically summarizes and outlines the main characteristics of Co-WPCs, focusing on their mechanical properties, dimensional stability, flame resistance, and weather resistance.

### 3.1 Mechanical properties of Co-WPCs

Since the core and shell layers of Co-WPCs control their mechanical properties, it is convenient to tailor the mechanical performance of Co-WPCs by adjusting the thickness of the shell layer and the quality of the core and shell layers [12, 17]. By filling the core and shell layers with varied amounts of wood flour, reinforced fibers, inorganic particles, and hybrid fillers, or by employing recycled thermoplastics as matrices, different characteristics of the layers can be achieved [12, 17–20]. Additionally, by applying an external force to the Co-WPCs, the shell layer can limit the deformation of the core layer and stop surface cracks from spreading to the center of the composites. Therefore, when Co-WPCs are subjected to external stress, a stronger and more flexible shell layer would aid in preserving their integrity and minimizing damage. Co-WPCs with a core-shell structure exhibit special mechanical qualities that depend on several variables.

The shell layer feeding rate and extrusion speed can be controlled to alter the shell thickness, which can then be used to alter the ratio of shell to core materials in Co-WPCs. When the wood flour loading in the shell layer was fixed, Yao and Wu [11] discovered that Co-WPCs with thicker shell layers exhibited higher impact strength and a lower modulus. When Jin and Matuana [12] co-extruded WPCs with PVC cap layers, the thickness of the PVC cap layers had a negative impact on the flexural modulus but a negligible influence on the flexural strength of the Co-WPCs. Co-Kim et al. [17] prepared Co-WPCs with shell layers that varied in thickness and core layers that were classified as weak, moderate, and strong. The relative magnitudes of the modulus and strength of the shell and core layers determined the variation in the modulus and strength of the Co-WPCs with different shell layer thicknesses. When the shell modulus and strength were greater than the core layer, a thicker shell enhanced the Co-WPCs' modulus and strength; conversely, when the core layer's characteristics were superior, the opposite was noted. When the shell thickness was increased for the weak, moderate, and strong core systems, the impact strength of the Co-WPCs rose as well. Co-WPCs' shell layer was typically filled with less material than their core layer when the preparation process and composite performance were taken into account. The impact properties of Co-WPCs are typically enhanced in low-filled shell layers, leading to lower strength and modulus. Increasing the thickness of the shell layer has been observed to improve impact performance while reducing modulus and strength. However, the strength of Co-WPCs may be increased compared to core-only materials in certain cases due to the fully enclosed structure. Ultimately, the impact of shell thickness on Co-WPCs is influenced by the properties of the shell layer. Enhancing the thickness of a high-performance shell layer is advantageous for improving the mechanical properties of Co-WPCs. Typically, the thickness of the shell layer is less than a quarter of the core thickness, and for reasons related to cost and protective efficiency, a thinner shell layer is recommended.

The core layer is a significant component of Co-WPCs, serving as their fundamental and primary element. By adjusting the composition of high-density polyethylene (HDPE) to low-density polyethylene (LDPE), modifying process parameters such as temperature and wood flour content, incorporating recycled polymers, and introducing materials like glass fiber, the qualities of the core layer can be customized. Research indicates that Co-WPCs exhibit superior mechanical properties when equipped with a high-quality core, even with identical shell layers. In contrast to core-only materials, the enhancement in Co-WPCs is more pronounced with a low-quality core; therefore, Co-WPCs featuring satisfactory mechanical characteristics can be achieved by incorporating a weaker core through suitable core-shell designs. Instead of utilizing WPCs as core layers, the integration of laminated veneer lumber (LVL) with exceptional mechanical properties results in the production of high-strength WPCs-LVL co-extruded composites. The high strength and minimal deformation of WPCs-LVL make them ideal materials for structural applications.

The constituents of the shell layer are crucial in determining the collective mechanical performance of Co-WPCs, with the ability to attain desirable mechanical properties through adjustments in the shell layer quality [14]. Variations in the shell layer quality were achieved by altering wood flour content [11, 12, 21], incorporating reinforced fibers [22, 23], introducing inorganic particles [9, 24–26], and utilizing hybrid fillers [18]. An overview of the primary fillers employed in the shell layers of Co-WPCs can be found in **Table 1**.

### 3.2 Dimensional stability of Co-WPCs

Wood materials have a high polarity, leading to significant dimensional changes upon moisture absorption, but minimal impacts from temperature fluctuations. In contrast, polymers typically absorb moisture at less than 1%, resulting in negligible effects on dimensional changes due to humidity, but considerable size deformations with temperature variations. Furthermore, polymers can exhibit creep deformation under prolonged external forces or gravity. WPCs are composed of polar wood fibers and non-polar polymers; their dimensional stability includes moisture-induced expansion and contraction, thermal expansion, and contraction due to temperature fluctuations, as well as creep deformation under long-term loading. Considering the diverse factors of humidity, temperature, and load based on different regions and

| | | Mechanical performance | | | |
|---|---|---|---|---|---|
| **Fillers** | **Percentage (wt%)** | **Flexural strength** | **Flexural modulus** | **Impact strength** | **References** |
| Wood flour | 0–27.5 | ↑[a] | ↑ | - | [12] |
| | 0–25, 35 | ↑, ↓[b] | ↑, ↑ | ↓, ↓ | [11] |
| | 0–20 | ↑ | ↑ | ↓ | [20] |
| CNTs | 5 | ↑ | ↑ | — | [19] |
| SGFs | 0–40 | ↑ | ↑ | ↓ | [23] |
| BFs | 0–30 | ↑ | ↑ | ↓ | [2] |
| Talc | 0–50 | ↑ | ↑ | ↓ | [25] |
| $SiO_2$ | 0–20 | ↑ | ↑ | ↑ | [26] |
| $CaCO_3$ | 0–25 | ↑ | ↑ | ↓ | [24] |
| TPCC/WF | 12/0–15, 12/25 | ↑, ↓ | ↑, ↓ | ↓ | [18] |

[a] ↑ = *Increase.*
[b] ↓ = *Decrease.*

**Table 1.**
*Effect of shell fillers on the mechanical performance of Co-WPCs [27].*

seasons is essential to understand the comprehensive impact on the dimensional stability of WPCs. Given the unique multilayer structure of Co-WPCs, investigating their dimensional stability is particularly crucial.

### *3.2.1 Water absorption*

The presence of abundant hydroxyl groups and the distinctive structure of wood flour render it hydrophilic. Wood-plastic composites (WPCs) combine hydrophilic wood flour with hydrophobic plastics, with the assumption that the wood flour is enclosed within the plastic matrix. However, traditional highly filled WPCs often encounter challenges in fully encapsulating the wood flour, allowing moisture to penetrate the composite easily. When water is absorbed, the wood flour swells, leading to the disruption of the interface between the wood flour and the matrix. Subsequently, microcracks may form as the water evaporates (as depicted in **Figure 4**), resulting in diminished mechanical properties of the WPCs and permitting further water ingress [5, 28].

Applying a pure hydrophobic polymer coating to WPCs can significantly enhance their resistance to moisture [29]. Jin and Matuana [12] utilized coextrusion to encapsulate WPCs with a PVC shell and studied the water absorption characteristics of the resulting composites. The inclusion of shell layers led to a reduction in both the rate and overall moisture content absorbed by the WPCs. While the introduction of wood flour in the shell layer increased the water absorption of Co-WPCs, the thickness of the shell layer itself did not impact their water absorption properties. This correlation was also found by Mei et al. [30]. However, Hao et al. [20] observed that the addition of wood flour content ranging from 0 to 20 wt% in the shell layer did not significantly alter the water absorption or thickness of Co-WPCs. However, Co-WPCs containing

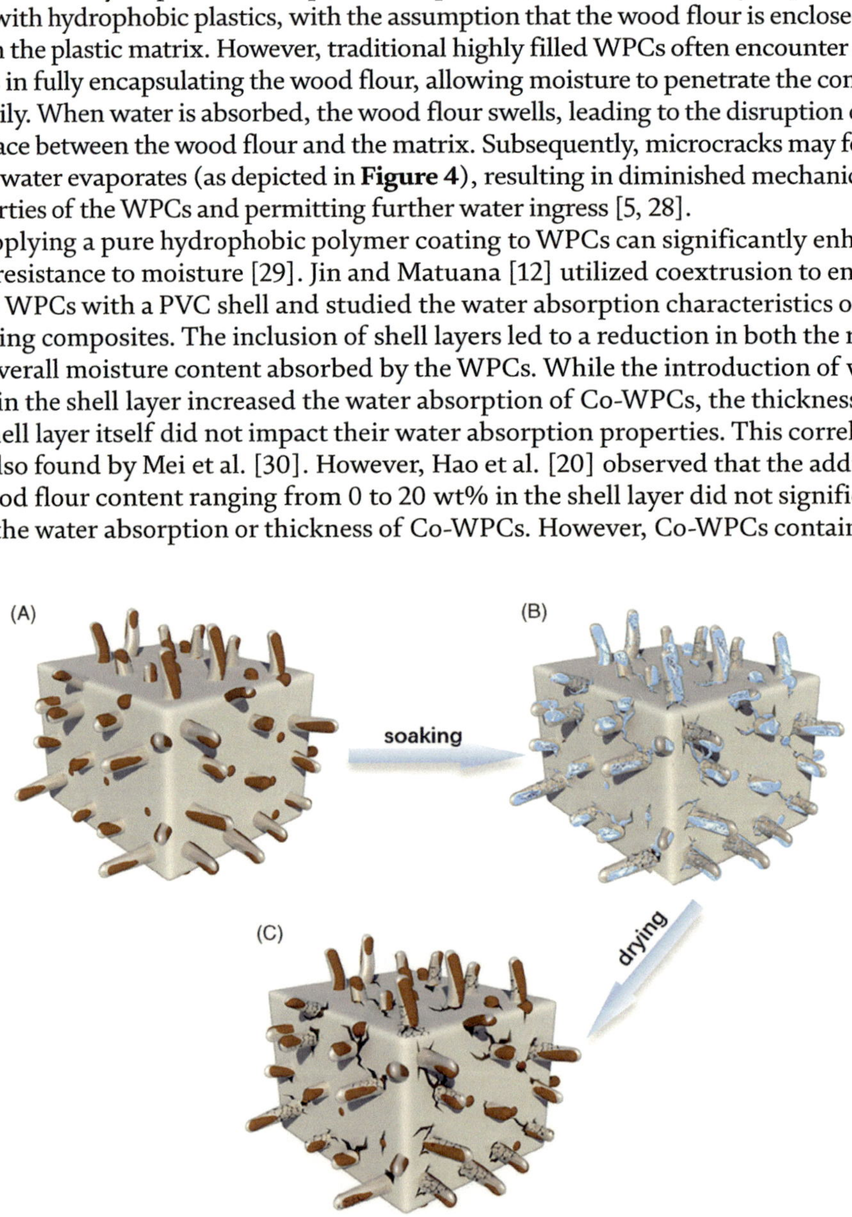


**Figure 4.**
*Schematic diagram of the water-induced failure of WPCs. (A) WPCs with wood flour capped well by a plastic matrix (B). Soaked WPCs with water in the interface of the wood flour and matrix (C). Dried WPCs with a broken wood flour/matrix interface [5].*

20 wt% wood flour still exhibited higher water absorption compared to other Co-WPCs. In a separate study, Kim et al. [18] developed Co-WPCs with shell layers filled with treated precipitated calcium carbonate (TPCC) and wood fiber (WF). The water absorption of these Co-WPCs was unaffected by the TPCC content in the shell layer but showed a direct relationship with the WF content. This was attributed to the effective dispersion of TPCC, enhanced interactions between TPCC and HDPE in the shell, and the hydrophilic nature of WF.

Turku et al. [8] incorporated various fillers, including aluminum trihydroxide (ATH), zinc borate (ZB), melamine, graphite (G), and titanium dioxide ($TiO_2$), into the shell of polypropylene (PP)-based Co-WPCs. The hydrophilic ZB filler increased the water absorption of Co-WPCs, while the other fillers resulted in reduced wettability of the composite. Additionally, the group introduced pulp cellulose (PC), a blend of PC/MFC (microfibrillated cellulose), and wood flour of different sizes into the shell of Co-WPCs. They discovered that Co-WPCs loaded with wood flour exhibited the highest water absorption and thickness swelling, whereas PC-filled Co-WPCs demonstrated lower moisture absorption. MFC-loaded Co-WPCs showed a similar water absorption trend as PC-filled Co-WPCs but a higher tendency for thickness swelling. The higher crystallinity of PC and MFC compared to wood flour, along with the impermeability of cellulose crystals to water, likely contributed to the reduced moisture absorption of PC- and MFC-filled Co-WPCs. The impact of incorporating graphite (G), expandable graphite (EG), carbon nanotubes (CNTs), carbon black (CB), and carbon fibers (CFs) into the shell of Co-WPCs on water absorption performance was also studied [22]. Co-WPCs containing EG exhibited the highest water absorption due to the lamellar and expanded structure of EG particles, while other fillers did not negatively affect the wettability of the composite. Furthermore, Butylina et al. [31] added $Fe_3O_4$, $TiO_2$, and zinc oxide (ZnO) to the Co-WPC shell and analyzed the wettability of the composites. The inclusion of $TiO_2$ increased water absorption by 43%, whereas iron oxide and ZnO did not impact composite wettability. However, $TiOO_2$-filled Co-WPCs showed minimal thickness swelling due to their low density and high porosity.

Typically, moisture absorption tests on Co-WPCs are conducted at 20°C, despite WPCs being exposed to temperatures as high as around 60°C during midday. In a study by Zhao et al. [5], the water absorption performance of Co-WPCs enclosed in an HDPE shell was compared to that of uncoextruded WPCs at 60°C. The Co-WPC exhibited reduced water absorption and enhanced dimensional stability compared to the uncoextruded WPC. Additionally, the Co-WPC maintained higher flexural properties than the uncoextruded WPC. Therefore, applying a hydrophobic shell layer through coextrusion onto WPCs can offer protection to the composites in environments with high temperatures and humidity.

### *3.2.2 Creep properties*

During long-term use, the creep behavior resulting from the movement and slippage of polymer matrix molecular chains in WPCs can significantly impact their macroscopic properties and service life. Given the higher polymer content in the shell layer, Co-WPCs exhibit more pronounced time- and temperature-dependent creep behavior during extended use [2, 20, 26]. Increasing the wood fiber content can effectively reduce the creep deformation of WPCs; studies have shown that increasing the wood fiber content (by mass fraction) from 50–70% can reduce the 24-hour creep deformation by 58% [32]. Similarly to mechanical properties, incorporating rigid particles such as silica or wood fibers in the shell layer can significantly reduce the creep deformation of the shell layer, aiming to decrease the overall long-term creep deformation of Co-WPCs. This is primarily attributed to the ability of rigid silica or wood fibers to effectively inhibit polymer chain slippage and rearrangement, thus enhancing their creep resistance [33, 34]. Compared to micron-sized silica, nano-sized silica can achieve a higher reduction in the creep deformation of the shell layer and Co-WPCs, mainly due to the reinforcing effect of nano-sized particles. Surface modification of nano-sized silica with silane can further enhance the creep resistance of Co-WPCs. However, while shell reinforcement has a limited impact on the overall creep resistance of Co-WPCs, increasing the wood fiber content in the core layer, which is the main component of Co-WPCs, can also effectively reduce the overall creep deformation of Co-WPCs.

#### *3.2.3 Thermal expansion properties*

When the temperature rises, materials frequently expand thermally because their free volume increases. The slope of the linear part of the dimensional change-temperature curve served as a proxy for the linear coefficient of thermal expansion (LCTE) of composites [35]. The depth of the atomic bond energy function and the reinforcing capacity of fillers in composites are connected with LCTE. A lower LCTE value is preferred for composite structural applications in order to guarantee thermal dimensional stability [36, 37]. The linear coefficient of thermal expansion of polymers is generally higher than 1 × 10–4°C-1, significantly exceeding those of metals and ceramics at 2 × 10–5°C-1. In contrast, the thermal expansion coefficient of natural biomass materials ranges from 5 × 10–6 to 5 × 10–5°C-1 [38, 39]. Therefore, the presence of wood fibers can effectively reduce the thermal expansion behavior of the polymer matrix. The addition of glass fibers [40], talc, basalt fibers (BFs), TPCC/wood flour, and talc/BFs to the shell layers of Co-WPCs all resulted in a reduction in the LCTE of the composites.

The complete encapsulation of the core layer by the shell layer structurally also to some extent restrains the thermal expansion behavior of Co-WPCs. Nevertheless, a difference in the linear coefficient of thermal expansion (LCTE) between the core and shell layers can lead to stress accumulation at their interface, exacerbated by higher temperatures [35]. This accumulated stress has the potential to disrupt the core-shell structure of Co-WPCs, ultimately leading to decreased mechanical properties, water absorption performance, flame resistance, and weathering characteristics. Hence, achieving a similar LCTE between the shell and core layers is advantageous for Co-WPCs. Additionally, the extrusion process of WPCs results in significant thermal expansion anisotropy due to the oriented arrangement of wood fibers. This means that the coefficient of thermal expansion (CTE) along the direction of the oriented wood fibers is much smaller than those in the thickness and other directions [41]. This indicates that wood fibers with different aspect ratios achieve a reduction in the coefficient of thermal expansion of Co-WPCs by limiting the deformation of the polymer matrix.

### 3.3 Weather resistance

As the use of WPCs in outdoor settings rises, they face exposure to factors like ultraviolet light, moisture, fungal growth, heat, and other environmental stressors. The yearly costs associated with the degradation of WPCs amount to millions of dollars [42]. Enhancing the weathering resistance of WPCs is crucial for their advancement and broadening their range of applications.

### *3.3.1 Accelerated ultraviolet weathering*

Typically, accelerated ultraviolet weathering tests involve exposing materials to several cycles of UV radiation, water spraying, and condensation to replicate sunlight, rain, and dew conditions [4]. These tests are designed to imitate the natural outdoor weathering process of WPCs over an extended period in a shorter timeframe within a laboratory setting. When exposed to outdoor elements, WPCs may experience color fading due to natural weathering, leading to a significant decline in both the visual appeal and functional longevity of the material. Analysis of color involves assessing attributes, such as lightness (L*), the red-green coordinate (a*), and the yellow-blue coordinate (b*) [29]. Color changes ($\Delta E_{ab}$) are typically determined using the following formula:

$$\Delta E_{ab} = \left(\Delta L^2 + \Delta a^2 + \Delta b^2\right)^{1/2} \quad (1)$$

In Eq. (1), ΔL, Δa, and Δb represent the variances in lightness (L*) and chromaticity coordinates (a* and b*) observed before and after the weathering process.

After 1000 hours of weathering, Co-WPCs with an HDPE cap exhibited less color change compared to uncapped WPCs, while Co-WPCs with a PP cap experienced greater color alteration than uncapped WPCs. The susceptibility of PP to UV degradation surpasses that of HDPE, leading to more pronounced discoloration in Co-WPCs with a PP cap. Matuana et al. [4] delved further into the degradation mechanism of HDPE-coated Co-WPCs. Photooxidation altered the surface chemistry of uncapped WPCs, resulting in darkening of the composites. Subsequently, the loss of pigmented wood components and increased surface roughness contributed to a lighter appearance of uncapped WPCs. Conversely, in Co-WPCs, the hydrophobic HDPE shell layer created through coextrusion prevented the loss of pigmented components, thereby significantly mitigating discoloration (as illustrated in **Figure 5**). The discoloration observed in Co-WPCs mainly stemmed from the photooxidation of wood constituents at the core-shell interface. Applying a transparent plastic shell layer via coextrusion on WPCs can serve as a safeguard against discoloration caused by weathering. The integration of UV absorbers or photostabilizers into the shell layers could offer added protection, potentially minimizing discoloration in Co-WPCs (as outlined

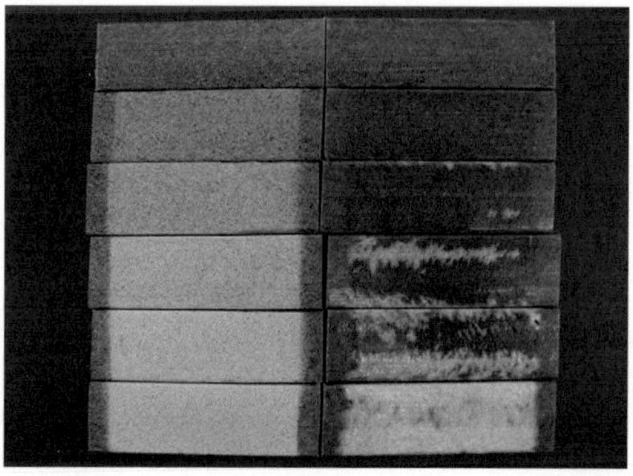

**Figure 5.**
*Discoloration of uncapped WPCs (left column) and Co-WPCs (right column) after various weathering times (0, 192, 432, 744, 1392, and 1952 h) [4].*

| Additives | Basic composition | Percentage (wt%) | ΔL | $\Delta E_{ab}$ | References |
|---|---|---|---|---|---|
| $TiO_2$ | HDPE/Compatibilizer | 1 | [a]↓ | ↓ | [29] |
| | PP/cellulose pulp | 3 | ↓ | ↓ | [31, 43] |
| | PP/cellulose pulp | 10 | | | [32] |
| Photostabilizer | HDPE/Compatibilizer | 0.6 | ↓ | ↓ | [29] |
| $Fe_3O_4$ | PP/cellulose pulp | 3 | ↓ | ↓ | [31, 43] |
| ZnO | PP/cellulose pulp | 3 | [b]↑ | ↑ | [31, 43] |
| Melamine | PP/cellulose pulp | 10 | ↓ | | [44] |
| Graphite- | PP/cellulose pulp | 10 | ↓ | | [44] |
| ATH | PP/cellulose pulp | 10 | ↓ | | [44] |
| ZB | PP/cellulose pulp | 10 | ↓ | | [44] |
| CB | PP/cellulose pulp | 3 | ↓ | | [45] |
| EG | PP/cellulose pulp | 3 | ↓ | | [45] |
| CNTs | PP/cellulose pulp | 3 | ↓ | | [45] |
| CFs | PP/cellulose pulp | 3 | ↓ | | [45] |

[a]↓ = *Decrease.*
[b]↑ = *Increase.*

**Table 2.**
*Summary of shell layer additives and their effects on the color protection of Co-WPCs [27].*

in **Table 2**) [29, 31, 42–45]. Additionally, adding metal oxide-based pigments (such as titanium dioxide, iron oxide, zinc oxide, etc.) to the shell layer can enhance color stability while improving mechanical performance [31, 43].

### *3.3.2 Freeze-thaw weathering*

With the expanding range of applications for WPCs, it is essential to consider the impact of extreme conditions on their lifespan. In regions with high latitudes, WPCs are subjected to UV light degradation as well as the effects of freeze-thaw cycles [46]. Exposure to lower temperatures can lead to WPCs becoming brittle, while increased moisture uptake can compromise interfacial adhesion, resulting in a significant loss of mechanical properties [46, 47]. Freeze-thaw weathering tests involve subjecting samples to cycles of immersion in water at 23°C for 70 ± 1 hours, freezing at −20°C for 24 hours, and drying at 70°C for 70 ± 1 hours. Turku and Kärki [48] examined the impact of incorporating various fire retardants (melamine, aluminum trihydrate, graphite, zinc borate, and TiO2) at a 10% concentration into the shell of Co-WPCs after five freeze-thaw cycles. They observed surface cracks and reduced mechanical properties in all composites, indicating that the fire retardants had minimal effect on the freeze-thaw resistance of Co-WPCs. Furthermore, melamine and zinc borate displayed instability in Co-WPCs, leading to partial leaching from the surface. The researchers then introduced different carbon fillers, including carbon black, graphite, expandable graphite, carbon nanotubes, and carbon fibers, at a 3% concentration into the shell of Co-WPCs. They found that Co-WPCs filled with carbon black, carbon nanotubes, and carbon fibers maintained a significant portion of their mechanical performance. Furthermore, carbon black and carbon nanotubes were effective reinforcements for Co-WPCs when compared to carbon fibers.

### 3.4 Fire resistance of Co-WPCs

The flammability of WPCs presents safety concerns that restrict their potential applications. Conventional methods for enhancing composite fire resistance involve incorporating fire retardants into the bulk materials. Cone calorimetry is commonly employed to evaluate flame-retardant properties and determine parameters, such as heat release rate (HRR), peak heat release rate (PHRR), ignition time (IT), total heat release (THR), total smoke release (TSR), average specific extinction area (ASEA), average carbon monoxide (CO) and carbon dioxide (CO2) values, average effective heat of combustion (AEHC), and average mass loss rate (AMLR), to evaluate the fire performance of composites. The peak heat release rate (PHRR) is particularly significant as it indicates fire intensity and growth rate.

Initially, the proximity of the heat source to the WPC surface rendered the internal fire retardants ineffective in preventing combustion. Additionally, the high concentration of fire retardants added to WPCs adversely impacted their mechanical properties [8]. Incorporating a fire-retardant shell layer through coextrusion onto WPCs can prevent combustion without the need for excessive fire-retardant content. Conversely, a pure plastic shell layer negatively influenced the flammability of Co-WPCs, as the peak and heat release rates of plastics surpassed those of wood flour [49–51]. Consequently, Co-WPCs with pure plastic shell layers were not recommended due to fire safety concerns. Wu et al. [2] implemented basalt fibers at 10%, 20%, and 30% weight content in the shell of Co-WPCs, with findings revealing a reduction in PHRR, THR, HRR, ASEA, AEHC, and AMLR as basalt fiber content increased. Compared to Co-WPCs with a pure HDPE shell, the 30% basalt fiber-loaded variant exhibited a 19.7% decrease in PHRR. The improved fire resistance of basalt fiber-filled Co-WPCs was attributed to incomplete combustion within the core-shell composites, although these materials remained more flammable than core-only composites. Another study by Huang et al. introduced talc at weight percentages of 5%, 15%, 25%, 35%, and 50% into the HDPE shell layer of Co-WPCs, noting enhanced flame resistance at higher talc loadings (35% and 50%). Only the 50% talc-filled Co-WPCs exhibited comparable flammability to core-only WPCs loaded with 55% wood flour (**Table 3**).

Turku et al. [8] conducted a study examining the impact of incorporating five fire retardants—aluminum trihydroxide (ATH), zinc borate (ZB), melamine, graphite, and titanium oxide ($TiO_2$)—into the shell of PP-based Co-WPCs on the flammability characteristics of the composite materials. The addition of fire retardants to Co-WPCs led to enhancements in peak heat release rate (PHRR), ignition time (IT), and average mass loss rate (AMLR) compared to unfilled Co-WPCs. Among the formulations tested, Co-WPCs with melamine exhibited the lowest PHRR, while the inclusion of graphite resulted in the most significant decrease in IT (indicating the longest ignition time). Building upon these findings, Nikolaeva and Kärki [52] introduced ammonium polyphosphate (APP), natural graphite, expandable graphite, and carbon nanotubes (CNTs) individually into the shell of melamine-filled Co-WPCs. Notably, Co-WPCs containing 20 wt.% melamine and 20 wt.% natural graphite demonstrated the most substantial improvement in flammability performance. The incorporation of 20 wt.% melamine and 20 wt.% APP also contributed to enhancements in PHRR, IT, total heat release (THR), and mass loss rate (MLR) of the Co-WPCs, although it notably increased smoke production and CO emissions. Interestingly, the APP-filled Co-WPCs exhibited a lower PHRR compared to melamine-filled counterparts, yet displaying the shortest ignition time (IT) [53]. To address this issue and enhance the

| Fire retardants | Percentage (wt%) | HRR ($kW/m^2$) | PHRR ($kW/m^2$) | IT (s) | THR ($MJ/m^2$) | TSR ($m^2/m^2$) | ASEA ($m^2/kg$) | AEHC (MJ/kg) | AMLR (g/s) | References |
|---|---|---|---|---|---|---|---|---|---|---|
| BF | 0–30 | ↓[a] | ↓ | | ↓ | | ↓ | ↓ | ↓ | [2] |
| Talc | 0–5, 15–50 | ↑[b], ↓ | ↑, ↓ | | ↑, ↓ | ↑, ↓ | | ↑ | | [49] |
| Graphite | 10 | | ↓ | ↑ | — | | | ↑ | ↓ | [8] |
| $TiO_2$ | 10 | | ↓ | ↑ | ↓ | | | ↑ | ↓ | [8] |
| Melamine/APP | 10/10 | ↓ | | - | - | ↑ | ↑ | | ↑ | [52] |
| | 20/20 | ↓ | | ↑ | ↓ | ↑ | ↑ | | ↑ | |
| Melamine/CNTs | 10/2 | ↓ | | ↓ | ↓ | ↓ | ↓ | | ↑ | [52] |
| Melamine/Natural graphite | 10/10 | ↓ | | ↑ | - | ↓ | ↓ | | ↑ | [52] |
| | 10/20 | ↓ | | ↑ | ↓ | ↓ | ↓ | | ↑ | |
| | 20/20 | ↓ | | ↑ | ↓ | ↓ | ↓ | | ↑ | |
| Melamine/Expandable graphite | 10/5 | — | | ↓ | — | ↓ | ↓ | | ↑ | [52] |
| ATH | 10 | ↓ | ↓ | ↑ | ↓ | ↓ | ↑ | ↑ | ↓ | [8, 53] |
| Melamine | 10 | ↓ | ↓ | ↑ | ↓ | ↓ | ↓ | ↑ | ↓ | [8, 53] |
| ZB | 10 | ↓ | ↓ | — | ↓ | ↑ | ↑ | ↓ | ↓ | [8, 53] |
| APP | 10 | ↓ | | ↓ | ↓ | ↓ | ↑ | | | [53] |
| Natural graphite | 10 | ↓ | | ↑ | ↓ | ↓ | ↑ | | | [53] |
| APP/Natural graphite | 5/5 | ↓ | | ↑ | ↓ | ↓ | ↑ | | | [53] |
| APP/Expandable graphite | 5/5 | ↓ | | ↑ | ↓ | ↓ | ↑ | | | [53] |

**Table 3.**
*Flammability properties of Co-WPCs after adding flame retardants [27].*

ignition time of the APP-filled Co-WPCs, natural graphite was incorporated into the formulation, as standalone natural graphite exhibited the longest IT. The combination of APP and natural graphite effectively improved the flame-retardant properties of the Co-WPCs, synergistically enhancing flame resistance. Although natural graphite did not independently enhance flame resistance in the Co-WPCs, it notably affected the ignition time in isolation. These results underscore the synergistic flame resistance benefits achieved by incorporating a blend of different fire retardants. Notably, the studies mentioned did not compare the flammability performance of the Co-WPCs with core-only WPCs.

## 4. Applications of Co-WPCs

After years of development, WPCs in Europe have reached a mature stage in the market, with outdoor flooring and automotive interiors accounting for about 90% of the total market share. While outdoor wall panels, fences, furniture, and other aspects hold a relatively small market share, they have shown strong growth momentum. In China, WPCs mainly include doors and windows, interior wall panels, and outdoor pallets. WPC product series are widely used in the construction industry, with many traditional building facilities made of wood and steel gradually being replaced by WPCs, such as decking, railings, fences, window frames, and exterior wall cladding. However, when used outdoors, issues such as moisture absorption, aging and fading, cracking and deformation, insect infestation, and mold growth caused by external factors such as water and sunlight remain important factors limiting the outdoor service life of ordinary WPC products. The use of a functionally co-extruded shell layer can effectively address and prevent the aforementioned problems.

### 4.1 Formworks

Formwork is a temporary structure used in concrete construction, serving as a critical tool for the formation of concrete structures. It consists primarily of panels, supporting structures, and connectors. The panels act as load-bearing plates directly in contact with newly poured concrete; the supporting structure serves as a temporary framework supporting the panels, concrete, and construction loads; and the connectors, as accessories, join the panels and supporting structure into a cohesive unit. Formwork must be constructed according to design specifications to ensure that concrete structures and components are formed in the prescribed positions and geometric dimensions, maintaining their correct alignment, and capable of withstanding their own weight as well as external loads acting upon them.

The commonly used materials for formwork in construction are plastic, plywood, and steel. However, plastic formwork has low rigidity and a high coefficient of thermal expansion; steel formwork is heavy, difficult to dismantle, prone to rust, and has high usage costs; plywood has a low number of uses and leads to significant wood wastage. In comparison to these materials, wood-plastic composite materials used for formwork offer the following advantages: high strength, good abrasion resistance, and physical and mechanical properties such as compression and impact resistance comparable to hardwood; high dimensional stability and good surface quality that remains flat even after multiple uses; low surface polarity, preventing concrete adhesion and facilitating easy form removal; resistance to aging and corrosion, insect-proof, impermeable to water, mud, and oil during casting, thus improving concrete quality; made from recycled waste plastic and agricultural and forestry residues, environmentally friendly, and cost-effective; favorable social and economic characteristics, aligning with sustainable development strategies and effectively alleviating the scarcity of resources such as steel and wood; excellent processing properties, allowing for multiple sawing, planning, and bonding processes, convenient fixing, and easy maintenance; lower density compared to steel, facilitating transportation and handling. As a result, the use of wood-plastic composite materials for formwork has gained increasing attention in recent years.

## 4.2 Architectural decoration field

### *4.2.1 Wallboard*

In architectural decoration materials, wall panels include interior wall protective panels and exterior wall decorative panels that possess excellent noise reduction and thermal insulation characteristics. They not only effectively maintain the surface of the wall but also serve as decorative elements. Compared to wall panels made of other materials, co-extruded wood-plastic wall panels offer advantages, such as anti-slip properties, durability, crack resistance, waterproofing, termite resistance, corrosion resistance, non-polluting, insulating, heat-insulating, flame-retardant, and easy maintenance. In recent years, co-extruded wood-plastic wall panels have become increasingly popular and are widely used in daily life. For instance, co-extruded wood-plastic wall panels can be used in modular kitchens, wall panels, bathroom cabinets, exterior wall claddings, prefabricated houses, signage, and display boards.

The decorative function of co-extruded wood-plastic wall panels can provide architectural and esthetic value, significantly enhancing the utilitarian value of building structures. Through coextrusion technology, wall panels with different colors, textures, and functionalities can be easily produced. Co-extruded wall panels can also be designed in various shapes, such as flat panels, long panels, circular panels, circular flat panels, and wave panels, meeting the demand for diverse decorative styles.

### *4.2.2 Ceiling*

In recent years, with the rapid development of China's economy and the improvement of people's living standards, interior decoration of residences and public buildings has become a burgeoning industry. To enhance esthetics, suspended ceiling decoration for interior ceilings and related corridor ceilings during renovations is becoming increasingly common. Traditionally, the materials used to make ceiling panels include wood, gypsum board, plastic board, aluminum alloy board, among others. Each of these ceiling panels has significant drawbacks: wooden ceiling panels are prone to cracking and deformation over long-term use, and their production consumes a considerable amount of wood, posing challenges to ecological preservation; plastic ceiling panels are expensive, prone to degradation, discoloration, deformation, and have low strength and a short lifespan; aluminum alloy ceiling panels offer high strength and modulus, resist deformation during use, but are heavy, challenging to install, and lack esthetic appeal; gypsum ceiling panels have weak strength, low impact resistance, and being made of thermosetting gypsum, do not degrade, posing serious environmental pollution concerns when disposed. In recent years, wood-plastic composite materials made mainly from discarded plastics and agricultural and forestry waste have gained attention due to their low density, high strength, recyclability, and ease of processing, among other advantages. Co-extruded wood-plastic materials with superior comprehensive performance have also been widely applied in the ceiling field.

#### *4.2.3 Floor*

Co-extruded wood-plastic flooring not only retains the traditional advantages of mold resistance, insect resistance, and other physical properties of wood-plastic materials but also possesses wear resistance, scratch resistance, stain resistance, and weather resistance. Experimental data have shown that the wear resistance and scratch resistance of co-extruded wood-plastic flooring are more than five times stronger than first-generation wood-plastic composite, effectively preventing damage from scratches caused by hard objects. Additionally, the robust surface layer of wood-plastic co-extruded flooring can resist the penetration of colored liquids and oily liquids, making the surface very easy to clean. This surface layer also enhances the resistance of wood-plastic flooring to sunlight, rain, snow, acid rain, and seawater, giving the flooring an exceptionally long lifespan. Co-extruded wood-plastic flooring primarily comes in solid core, round hole, and square hole structures.

#### *4.2.4 Skirting line*

Baseboards are installed at the junction of the floor and the wall, serving to provide visual balance and enhancing the overall esthetic appeal of the interior space. Utilizing their linear design, material, and color, baseboards can complement and harmonize with the indoor environment, adding a decorative touch. Baseboards help strengthen the connection between the wall and the floor, reducing wall deformation and preventing damage from external impacts. Additionally, baseboards are easy to clean, making it convenient to wipe away dirty water or spills during floor cleaning. In addition to their protective function for the wall, baseboards play a significant role in enhancing the overall visual appeal of home decor. They act as the boundary line of the floor, often attracting the eye naturally, with a typical thickness of 5–10 mm when extended from the wall during installation. Co-extruded wood-plastic composite materials, which can be easily optimized for hardness and waterproofing, are also used in the production of baseboards.

### **4.3 Landscape architecture**

There are many forms of application for wood-plastic composite materials in paving design, such as garden pathways, courtyard paving, observation platforms, wooden boardwalks, and more. In comparison to standard wood-plastic composite materials, co-extruded wood-plastic flooring features an added protective layer, clear wood grain, and natural colors. Furthermore, the four-sided coextrusion fully protects the core layer of the board, making it more resistant to wear, scratches, and stains. In high-traffic areas, it effectively prevents scratches from foot traffic and damage from hard objects, enduring various harsh tests from nature. Additionally, co-extruded wood-plastic flooring offers users high practical value, esthetic enjoyment, and visual appeal.

In architecture, a wall refers to a vertical spatial partition structure used to enclose, divide, or protect a certain area. Nearly all significant building materials can be used for constructing walls: wood, stone, brick, concrete, metal, glass, and even vegetation. By utilizing wood-plastic composite materials in wall construction, the aging resistance of the walls can be greatly enhanced. Additionally, the repair methods for wood-plastic composite materials are relatively straightforward, which can significantly improve the practical performance of the walls. Therefore, applying wood-plastic composite materials in wall construction can enhance both the performance and esthetics of the walls. Co-extruded wood-plastic composite materials have advantages in weather resistance and visual appeal compared to standard wood-plastic composite materials, making them highly suitable for use in walls.

Railings are a key component in landscape construction, primarily used for personal safety and the protection of equipment, facilities, and surroundings. Their applications include lake-side railings, riverbank railings, mountain path railings, and road guardrails. Railings have high safety requirements, and wood-plastic composite materials possess strong impact resistance, ensuring the strength and stability of the railings. Furthermore, wood-plastic composite materials are highly malleable, meeting the esthetic needs of railings. They can be designed with different patterns, shapes, and colors according to the requirements of landscape design. As society increasingly agrees on the importance of ecological environment and forest conservation, high-end raw wood will be replaced by high-quality weather-resistant outdoor composite materials. Co-extruded wood-plastic composite materials not only have excellent weather resistance but also provide a lasting tactile experience, meeting the higher standards and requirements of various landscape projects to create a harmonious, peaceful, clear, and serene ecological landscape.

Pavilions in landscape design are often designated for relaxation, with most pavilions being wooden structures. By utilizing wood-plastic composite materials with strong malleability, exquisitely designed small pavilions can be created, incorporating light wood tones paired with natural textures to connect seamlessly with long leisure boardwalks, achieving eco-friendly innovative design. Furthermore, wood-plastic composite materials can provide a sense of comfort and tranquility in terms of auditory perception. From a scientific perspective, wood-plastic composite materials can absorb some of the more harsh sound waves, creating a quiet and comfortable ambiance when people converse in spaces constructed using wood-plastic composite materials.

Pergolas are important urban landscapes that can both separate and connect two distinct spaces. In modern society, some exquisitely designed pergolas are made of wood structures, and by using wood-plastic composite materials, the natural wood texture can be replicated. Pergolas constructed from co-extruded wood-plastic composite materials do not degrade or peel after long-term exposure to rain, exhibiting strong stability while maintaining the texture and appearance of wood, making them popular among people. Therefore, we can see wood-plastic pergolas being utilized in many residential areas and parks. For instance, in a lattice wood structure pergola, sunlight can filter through the structure, creating a warm and inviting atmosphere for those seated within.

By optimizing material and structural design, co-extruded wood-plastic composite materials have greatly enhanced weather resistance, stability, and a rich wood texture compared to standard wood-plastic composite materials. In landscape design, a large number of street furniture items, such as seats, benches, trash cans, flower boxes, tree pots, and walkways, are made from co-extruded wood-plastic composite materials.

## 5. Conclusions

Co-extruded WPCs (Co-WPCs) with core-shell structures are versatile and can be tailored to meet various requirements for different application scenarios. Common non-fully wrapped structures are not recommended, unless cost savings outweigh the loss in properties. Non-fully wrapped Co-WPCs with a fixed groove design or a sandwich structure are acceptable. Co-WPCs with a solid wood core and specially structured Co-WPCs are suitable for high-performance applications. Applying a protective outer layer to WPCs can effectively reduce moisture absorption and degradation caused by weathering. The incorporation of fillers into the outer layer has been found to enhance the mechanical properties, durability, flame resistance, and dimensional stability of Co-WPCs. However, optimizing the performance of Co-WPCs involves balancing various factors. For instance, a fully hydrophobic plastic coating can reduce water uptake but may compromise tensile and flexural strength. While fillers can improve dimensional stability, they may also complicate the manufacturing process. The enhanced weathering resistance of Co-WPCs may result in a darker appearance due to the inclusion of carbon fillers. Although most reinforced Co-WPCs exhibit increased tensile and flexural strength, their impact resistance tends to decrease. Co-WPCs have been successfully applied in formwork, architectural decoration field such as wallboard, ceiling, floor, and skirting line, and landscape architectures including outdoor flooring, wall, guardrail, pavilion, porch frame, and landscape pieces. Future research should focus on tailoring Co-WPCs to specific applications, such as enhancing flame resistance for indoor use, strengthening mechanical properties for construction applications, and increasing outdoor durability. The development of multifunctional fillers capable of improving multiple performance aspects of Co-WPCs is a key area for advancement. Co-WPCs are widely applied in daily life and are considered the most promising biomass composite material with great application potential.

## Acknowledgements

This work is supported by the following grants and programs: 1. Yunnan Agricultural Joint Research Key Project (202301BD070001-153); 2. Yunnan Provincial Applied and Basic Research Grants (202201AT070058); 3. National Natural Science Foundation of China (32060381); 4. The High Level Innovative One-Ten-Thousand Youth Talents of Yunnan Province (YNWR-QNBJ-2020-203); 5. “111” Project (D21027).

## Conflict of interest

The authors declare no conflict of interest.

## Author details

Kaimeng Xu[1], Huanbo Wang[1] and Tian Liu[2*]

1 Yunnan Provincial Key Laboratory of Wood Adhesives and Glued Products, International Joint Research Center for Biomass Materials, Southwest Forestry University, Kunming, China

2 Key Laboratory of Bio-based Material Science and Technology (Northeast Forestry University), Ministry of Education, Engineering Research Center of Advanced Wooden Materials, Harbin, China

*Address all correspondence to: granthill63@163.com

## References

[1] Teuber L, Osburg V-S, Toporowski W, Militz H, Krause A. Wood polymer composites and their contribution to cascading utilisation. Journal of Cleaner Production. 2016;**110**:9-15. DOI: 10.1016/j.jclepro.2015.04.009

[2] Wu Q, Chi K, Wu Y, Lee S. Mechanical, thermal expansion, and flammability properties of co-extruded wood polymer composites with basalt fiber reinforced shells. Materials & Design. 2014;**60**:334-342. DOI: 10.1016/j.matdes.2014.04.010

[3] Pickering KL, Efendy MGA, Le TM. A review of recent developments in natural fibre composites and their mechanical performance. Composites Part A: Applied Science and Manufacturing. 2016;**83**:98-112. DOI: 10.1016/j.compositesa.2015.08.038

[4] Matuana LM, Jin S, Stark NM. Ultraviolet weathering of HDPE/wood-flour composites coextruded with a clear HDPE cap layer. Polymer Degradation and Stability. 2011;**96**(1):97-106. DOI: 10.1016/j.polymdegradstab.2010.10.003

[5] Zhao Z, Zhang Q, Wang J. Properties comparison of coextruded and un-coextruded wood plastic composites under high temperature accelerated water uptake. Polymer Composites. 2020;**41**(7):2616-2624. DOI: 10.1002/pc.25560

[6] Clemons C, Ibach RE. Effects of processing method and moisture history on laboratory fungal resistance of wood-HDPE composites. Forest Products Society. 2004;**54**(4):50-57

[7] Stark NM, Matuana LM. Characterization of weathered wood–plastic composite surfaces using FTIR spectroscopy, contact angle, and XPS. Polymer Degradation and Stability. 2007;**92**(10):1883-1890. DOI: 10.1016/j.polymdegradstab. 2007.06.017

[8] Turku I, Nikolaeva M, Kärki T. The effect of fire retardants on the flammability, mechanical properties, and wettability of co-extruded PP-based wood-plastic composites. BioResources. 2014;**9**(1):1539-1551. DOI: 10.15376/biores.9.1.1539-1551

[9] Dooley J, Hyun KS, Hughes K. An experimental study on the effect of polymer viscoelasticity on layer rearrangement in coextruded structures. Polymer Engineering & Science. 1998;**38**(7):1060-1071. DOI: 10.1002/pen.10274

[10] Huang RZ, Zhou CJ, Zhang Y, Wu QL. Co-extrusion Technology for Functioned Nature Fiber Reinforced Polymer Composites. Advanced Materials Research. 2013;**773**:497-501. DOI: 10.4028/www.scientific.net/AMR.773.497

[11] Yao F, Wu Q. Cocxtruded polyethylene and wood-flour composite: Effect of shell thickness, wood loading, and core quality. Journal of Applied Polymer Science. 2010;**118**(6):3594-3601. DOI: 10.1002/app.32742

[12] Jin S, Matuana LM. Coextruded PVC/wood-flour composites with WPC cap layers. Journal of Vinyl and Additive Technology. 2008;**14**(4):197-203. DOI: 10.1002/vnl.20162

[13] Hao X-L, Zhou H-Y, Sun L-C, Lin D-R, Ou R-X, Wang Q-W. Research progress and application of co-extruded wood plastic composites. Journal of

Forestry Engineering. 2021;**6**(5):27-38. DOI: 10.5555/20220235273

[14] Xiong W, Wu Q, Cai C. Mechanical and thermal performance of coextruded wood plastic composites for structural applications. Advances in Structural Engineering. 2013;**16**(5): 909-929. DOI: 10.1260/1369-4332.16.5.909

[15] Zong G, Hao X, Hao J, Tang W, Fang Y, Ou R, et al. High-strength, lightweight, co-extruded wood flour-polyvinyl chloride/lumber composites: Effects of wood content in shell layer on mechanical properties, creep resistance, and dimensional stability. Journal of Cleaner Production. 2020;**244**:118860-118869. DOI: 10.1016/j.jclepro.2019.118860

[16] Zong G, Zhou J, Zhang M, Ma Y, Zhao Y, He X, et al. Effect of mortise and Tenon structure on the properties of wood flour polyvinyl chloride-laminated veneer lumber Co-extruded composites. Polymers. 2023;**15**(9):2151-2163. DOI: 10.3390/polym15092151

[17] Kim BJ, Huang R, Han J, Lee S, Wu Q. Mechanical and morphological properties of coextruded wood plastic composites with glass fiber-filled shell. Polymer Composites. 2014;**37**(3):824-834. DOI: 10.1002/pc.23240

[18] Kim B-J, Yao F, Han G, Wang Q, Wu Q. Mechanical and physical properties of core–shell structured wood plastic composites: Effect of shells with hybrid mineral and wood fillers. Composites Part B: Engineering. 2013;**45**(1):1040-1048. DOI: 10.1016/j.compositesb.2012.07.031

[19] Jin S, Matuana LM. Wood/plastic composites co-extruded with multi-walled carbon nanotube-filled rigid poly(vinyl chloride) cap layer. Polymer International. 2009;**59**(5):648-657. DOI: 10.1002/pi.2745

[20] Hao X, Yi X, Sun L, Tu D, Wang Q, Ou R. Mechanical properties, creep resistance, and dimensional stability of core/shell structured wood flour/polyethylene composites with highly filled core layer. Construction and Building Materials. 2019;**226**:879-887. DOI: 10.1016/j.conbuildmat.2019.07.329

[21] Migneault S, Koubaa A, Erchiqui F, Chaala A, Englund K, Wolcott MP. Effects of processing method and fiber size on the structure and properties of wood–plastic composites. Composites Part A: Applied Science and Manufacturing. 2009;**40**(1):80-85. DOI: 10.1016/j.compositesa.2008.10.004

[22] Turku I, Kärki T. The influence of different carbon type fillers on the mechanical and physical properties of Co-extruded PP-based WPC. Advanced Materials Research. 2014;**1025-1026**:200-207. DOI: 10.4028/www.scientific.net/AMR.1025-1026.200

[23] Kim BJ, Huang R, Han J, Lee S, Wu Q. Mechanical and morphological properties of coextruded wood plastic composites with glass fiber-filled shell. Polymer Composites. 2016;**37**(3):824-834. DOI: 10.1002/pc.23240

[24] Xian Y, Feng T, Li H, Xing Z, Wang G, Cheng H. The influence of Nano-CaCO3 on the mechanical and dynamic Thermo-mechanical properties of Core–Shell structured wood plastic composites. Forests. 2023;**14**(2):257-267. DOI: 10.3390/f14020257

[25] Huang R, Kim B-J, Lee S, Zhang Y, Wu Q. Co-extruded wood-plastic composites with talc-filled shells: Morphology, mechanical, and thermal expansion performance. BioResources. 2013;**8**(2):2283-2299. DOI: 10.15376/biores.8.2.2283-2299

[26] Hao X, Zhou H, Xie Y, Xiao Z, Wang H, Wang Q. Mechanical reinforcement and creep resistance of coextruded wood flour/polyethylene composites by shell-layer treatment with nano- and micro-SiO2 particles. Polymer Composites. 2019;**40**(4):1576-1584. DOI: 10.1002/pc.24901

[27] Wang H, Zhang X, Guo S, Liu T. A review of coextruded wood–plastic composites. Polymer Composites. 2021;**42**(9):4174-4186. DOI: 10.1002/pc.26189

[28] Tamrakar S, Lopez-Anido RA. Water absorption of wood polypropylene composite sheet piles and its influence on mechanical properties. Construction and Building Materials. 2011;**25**(10):3977-3988. DOI: 10.1016/j.conbuildmat.2011.04.031

[29] Stark NM, Matuana LM. Coating WPCs using co-extrusion to improve durability. In: Conference for Coating Wood and Wood Composites: Designing for Durability. Seattle, WA; 2007

[30] Mei C, Sun X, Wan M, Wu Q, Chun S-J, Lee S. Coextruded wood plastic composites containing recycled wood Fibers treated with micronized copper-Quat: Mechanical, moisture absorption, and chemical leaching performance. Waste and Biomass Valorization. 2017;**9**(11):2237-2244. DOI: 10.1007/s12649-017-9992-z

[31] Butylina S, Martikka O, Kärki T. Effect of inorganic pigments on the properties of coextruded polypropylene-based composites. Journal of Thermoplastic Composite Materials. 2016;**31**(1):23-33. DOI: 10.1177/0892705716646416

[32] Sykacek E, Hrabalova M, Frech H, Mundigler N. Extrusion of five biopolymers reinforced with increasing wood flour concentration on a production machine, injection moulding and mechanical performance. Composites Part A: Applied Science and Manufacturing. 2009;**40**(8):1272-1282. DOI: 10.1016/j.compositesa.2009.05.023

[33] Jia M, Xue P, Zhao Y, Wang K. Creep behaviour of wood flour/poly (vinyl chloride) composites. Journal of Wuhan University of Technology-Mater. Sci. Ed. 2009;**24**:440-447. DOI: 10.1007/s11595-009-3440-2

[34] Georgiopoulos P, Kontou E, Christopoulos A. Short-term creep behavior of a biodegradable polymer reinforced with wood-fibers. Composites Part B: Engineering. 2015;**80**:134-144. DOI: 10.1016/j.compositesb.2015.05.046

[35] Nakagaito AN, Yano H. The effect of fiber content on the mechanical and thermal expansion properties of biocomposites based on microfibrillated cellulose. Cellulose. 2008;**15**(4):555-559. DOI: 10.1007/s10570-008-9212-x

[36] Zhu RP, Sun CT. Effects of Fiber orientation and elastic constants on coefficients of thermal expansion in laminates. Mechanics of Advanced Materials and Structures. 2003;**10**(2):99-107. DOI: 10.1080/15376490306733

[37] Singh S, Mohanty A. Wood fiber reinforced bacterial bioplastic composites: Fabrication and performance evaluation. Composites Science and Technology. 2007;**67**(9):1753-1763. DOI: 10.1016/j.compscitech.2006.11.009

[38] Monroe J, Gehring D, Karaman I, Arroyave R, Brown DW, Clausen B. Tailored thermal expansion alloys. Acta Materialia. 2016;**102**:333-341. DOI: 10.1016/j.actamat.2015.09.012

[39] Takenaka K, Ichigo M. Thermal expansion adjustable polymer matrix

composites with giant negative thermal expansion filler. Composites Science and Technology. 2014;**104**:47-51. DOI: 10.1016/j.compscitech.2014.08.029

[40] Huang R, Xiong W, Xu X, Wu Q. Thermal expansion behavior of co-extruded wood-plastic composites with glass-fiber reinforced shells. BioResources. 2012;7(4):5514-5526. DOI: 10.15376/biores.7.4.5514-5526

[41] Hao X, Zhou H, Mu B, Chen L, Guo Q, Yi X, et al. Effects of fiber geometry and orientation distribution on the anisotropy of mechanical properties, creep behavior, and thermal expansion of natural fiber/HDPE composites. Composites Part B: Engineering. 2020;**185**:107778. DOI: 10.1016/j.compositesb.2020.107778

[42] Matuana LM, Kamdem DP. Accelerated ultraviolet weathering of PVC/wood-flour composites. Polymer Engineering & Science. 2002;**42**(8):1657-1666. DOI: 10.1002/pen.11060

[43] Butylina S, Martikka O, Kärki T. Weathering properties of coextruded polypropylene-based composites containing inorganic pigments. Polymer Degradation and Stability. 2015;**120**:10-16. DOI: 10.1016/j.polymdegradstab.2015.06.004

[44] Turku I, Kärki T. Accelerated weathering of fire-retarded wood–polypropylene composites. Composites Part A: Applied Science and Manufacturing. 2016;**81**:305-312. DOI: 10.1016/j.compositesa.2015.11.028

[45] Turku I, Kärki T. Accelerated weathering of wood–polypropylene composite containing carbon fillers. Journal of Composite Materials. 2015;**50**(10):1387-1393. DOI: 10.1177/0021998315591842

[46] Pilarski JM, Matuana LM. Durability of wood flour-plastic composites exposed to accelerated freeze-thaw cycling. Part I. Rigid PVC matrix. Journal of vinyl and additive. Technology. 2005;**11**(1):1-8. DOI: 10.1002/vnl.20029

[47] Pilarski JM, Matuana LM. Durability of wood flour-plastic composites exposed to accelerated freeze–thaw cycling. II. High density polyethylene matrix. Journal of Applied Polymer Science. 2005;**100**(1):35-39. DOI: 10.1002/app.22877

[48] Turku I, Kärki T. Durability of fire-retarded wood-polypropylene composites exposed to freeze-thaw cycling. Baltic Forestry. 2016;**22**(2):341-347

[49] Huang R, Zhang X, Chen Z, Wan M, Wu Q. Thermal stability and flame resistance of the coextruded wood-plastic composites containing talc-filled plastic shells. International Journal of Polymer Science. 2020;**2020**:1435249. DOI: 10.1155/2020/1435249

[50] Stark NM, White RH, Mueller SA, Osswald TA. Evaluation of various fire retardants for use in wood flour–polyethylene composites. Polymer Degradation and Stability. 2010;**95**(9):1903-1910. DOI: 10.1016/j.polymdegradstab.2010.04.014

[51] Arao Y, Nakamura S, Tomita Y, Takakuwa K, Umemura T, Tanaka T. Improvement on fire retardancy of wood flour/polypropylene composites using various fire retardants. Polymer Degradation and Stability. 2014;**100**:79-85. DOI: 10.1016/j.polymdegradstab.2013.12.022

[52] Nikolaeva M, Kärki T. Evaluation of different flame retardant combinations for Core/Shell structured wood-plastic

composites by using a cone calorimeter. Advanced Materials Research. 2015;**1120-1121**:535-544. DOI: 10.4028/www.scientific.net/AMR.1120-1121.535

[53] Nikolaeva M, Kärki T. Influence of fire retardants on the reaction-to-fire properties of coextruded wood–polypropylene composites. Fire and Materials. 2015;**40**(4):535-543. DOI: 10.1002/fam.2308